Universitext

Manfredo P. do Carmo

Differential Forms and Applications

With 18 Figures

Springer-Verlag
Berlin Heidelberg New York
London Paris Tokyo
Hong Kong Barcelona
Budapest

Manfredo P. do Carmo
Instituto de Matematica Pura e Aplicada (IMPA)
Estrada Dona Castorina, 110
22460-320 Rio de Janeiro
Brazil

This is a translation of the Portuguese book "Formas Diferenciais e Apliçōes", first published by IMPA in 1971.

Mathematics Subject Classification (1991):
53-01, 53A05, 58A10, 58Z05, 70Hxx

ISBN 3-540-57618-5 Springer-Verlag Berlin Heidelberg New York
ISBN 0-387-57618-5 Springer-Verlag New York Berlin Heidelberg

Library of Congress Cataloging-in-Publication Data. Carmo, ManfredoPerdigaõ do. [Formas diferenciais e aplicações. English] Differential forms and applications / Manfredo P. do Carmo. p. cm. -- (Universitext) Includes bibliographical references and index. ISBN 0-387-57618-5 1. Differential forms. I. Title. QA381.C2813 1994 515'.37--dc20 94-21965

Springer-Verlag Berlin Heidelberg New York
a member of BertelsmannSpringer Science+Business Media GmbH

Printed in Germany

The final art for the drawings was prepared by Manfredo do Carmo Junior.

Typesetting: Camera-ready by the author using a Springer TEX macro package
41/3111 - 5 4 - Printed on acid-free paper

To my friends around the world, without whose help neither this book nor its author would be seeing the light.

Preface

This is a free translation of a set of notes published originally in Portuguese in 1971. They were translated for a course in the College of Differential Geometry, ICTP, Trieste, 1989. In the English translation we omitted a chapter on the Frobenius theorem and an appendix on the nonexistence of a complete hyperbolic plane in euclidean 3-space (Hilbert's theorem). For the present edition, we introduced a chapter on line integrals.

In Chapter 1 we introduce the differential forms in R^n. We only assume an elementary knowledge of calculus, and the chapter can be used as a basis for a course on differential forms for "users" of Mathematics.

In Chapter 2 we start integrating differential forms of degree one along curves in R^n. This already allows some applications of the ideas of Chapter 1. This material is not used in the rest of the book.

In Chapter 3 we present the basic notions of differentiable manifolds. It is useful (but not essential) that the reader be familiar with the notion of a regular surface in R^3.

In Chapter 4 we introduce the notion of manifold with boundary and prove Stokes theorem and Poincare's lemma.

Starting from this basic material, we could follow any of the possible routes for applications: Topology, Differential Geometry, Mechanics, Lie Groups, etc. We have chosen Differential Geometry. For simplicity, we restricted ourselves to surfaces.

Thus in Chapter 5 we develop the method of moving frames of Elie Cartan for surfaces. We first treat immersed surfaces and next the intrinsic geometry of surfaces.

Finally, in Chapter 6, we prove the Gauss-Bonnet theorem for compact orientable surfaces. The proof we present here is essentially due to S.S.Chern. We also prove a relation, due to M. Morse, between the Euler characteristic of such a surface and the critical points of a certain class of differentiable functions on the surface.

As most authors, I am indebted to so many sources that it is hardly possible to acknowledge them all. Let me at least mention that the first four chapters were strongly influenced by the writings of my friend and colleague Elon Lima and the last two chapters bear the imprint of my teacher and friend S.S. Chern.

For the present version I am indebted to my colleagues M. Dajczer, L. Rodríguez and W. Santos for reading critically the manuscript and offering a number of useful suggestions. Special thanks are due to Lucio Rodríguez for his care in the camera ready presentation of the final text.

Rio de Janeiro, February 1994. Manfredo Perdigão do Carmo

Table of Contents

1. Differential Forms in $\mathbf{R}^n$

The goal of this chapter is to define in $\mathbf{R}^n$ "fields of alternate forms" that will be used later to obtain geometric results.

In order to fix the ideas, we will work initially with the three-dimensional space $\mathbf{R}^3$.

Let p be a point of $\mathbf{R}^3$. The set of vectors $q - p$, $q \in \mathbf{R}^3$ (that have origin at p) will be called the *tangent space of* $\mathbf{R}^3$ *at* p and will be denoted by $\mathbf{R}^3_p$. The vectors $e_1 = (1,0,0)$, $e_2 = (0,1,0)$, $e_3 = (0,0,1)$ of the canonical basis of $\mathbf{R}^3_0$ will be identified with their translates $(e_1)_p, (e_2)_p, (e_3)_p$ at the point p.

A *vector field* in $\mathbf{R}^3$ is a map v that associates to each point $p \in \mathbf{R}^3$ a vector $v(p) \in \mathbf{R}^3_p$. We can write v as

$$v(p) = a_1(p)e_1 + a_2(p)e_2 + a_3(p)e_3,$$

thereby defining three functions $a_i\colon \mathbf{R}^3 \to \mathbf{R}$, $i = 1,2,3$, that characterize the vector field v. We say that v is *differentiable* if the functions a_i are differentiable.

To each tangent space $\mathbf{R}^3_p$ we can associate its *dual space* $(\mathbf{R}^3_p)^*$ which is the set of linear maps $\varphi\colon \mathbf{R}^3_p \to \mathbf{R}$. A basis for $(\mathbf{R}^3_p)^*$ is obtained by taking $(dx_i)_p$, $i = 1,2,3$, where $x_i\colon \mathbf{R}^3 \to \mathbf{R}$ is the map which assigns to each point its ith -coordinate. The set

$$\{(dx_i)_p;\ i = 1,2,3\}$$

is in fact the dual basis of $\{(e_i)_p\}$ since

$$(dx_i)_p(e_j) = \frac{\partial x_i}{\partial x_j} = \begin{cases} 0, & \text{if } i \neq j \\ 1, & \text{if } i = j. \end{cases}$$

Definition 1. A *field of linear forms* (or an *exterior form of degree* 1) in $\mathbf{R}^3$ is a map ω that associates to each $p \in \mathbf{R}^3$ an element $\omega(p) \in (\mathbf{R}^3_p)^*$; ω can be written as

$$\omega(p) = a_1(p)(dx_1)_p + a_2(p)(dx_2)_p + a_3(p)(dx_3)_p$$

or

$$\omega = \sum_{i=1}^{3} a_i \, dx_i,$$

where a_i are real functions in $\mathbf{R}^3$. If the functions a_i are differentiable, ω is called a *differential form of degree* 1.

Now let $\Lambda^2(\mathbf{R}_p^3)^*$ be the set of maps $\varphi: \mathbf{R}_p^3 \times \mathbf{R}_p^3 \to \mathbf{R}$ that are bilinear (i.e., φ is linear in each variable) and alternate (i.e., $\varphi(v_1, v_2) = -\varphi(v_2, v_1)$). With the usual operations of functions, the set $\Lambda^2(\mathbf{R}_p^3)^*$ becomes a vector space.

When φ_1 and φ_2 belong to $(\mathbf{R}_p^3)^*$, we can obtain an element $\varphi_1 \wedge \varphi_2 \in \Lambda^2(\mathbf{R}_p^3)^*$ by setting

$$(\varphi_1 \wedge \varphi_2)(v_1, v_2) = \det(\varphi_i(v_j))$$

The element $(dx_i)_p \wedge (dx_j)_p \in \Lambda^2(\mathbf{R}_p^3)^*$ will be denoted by $(dx_i \wedge dx_j)_p$. It is easy to see that the set $\{(dx_i \wedge dx_j)_p, \ i < j\}$ is a basis for $\Lambda^2(\mathbf{R}_p^3)^*$ (this will be proved in a more general setting in Proposition 1 below). Furthermore,

$$(dx_i \wedge dx_j)_p = -(dx_j \wedge dx_i)_p, \qquad i \neq j,$$

and

$$(dx_i \wedge dx_i)_p = 0.$$

Definition 2. A *field of bilinear alternating forms* or *an exterior form of degree 2* in $\mathbf{R}^3$ is a correspondence w that associates to each $p \in \mathbf{R}^3$ an element $\omega(p) \in \Lambda^2(\mathbf{R}_p^3)^*$; ω can be written in the form

$$\omega(p) = a_{12}(p)(dx_1 \wedge dx_2)_p + a_{13}(p)(dx_1 \wedge dx_3)_p + a_{23}(p)(dx_2 \wedge dx_3)_p$$

or

$$\omega = \sum_{i<j} a_{ij} dx_i \wedge dx_j, \qquad i, j = 1, 2, 3,$$

where a_{ij} are real functions in $\mathbf{R}^3$. When the functions a_{ij} are differentiable, ω is a *differential form of degree* 2.

We will now generalize the notion of differential form to $\mathbf{R}^n$. Let $p \in \mathbf{R}^n$, $\mathbf{R}_p^n$ the tangent space of $\mathbf{R}^n$ at p and $(\mathbf{R}_p^n)^*$ its dual space. Let $\Lambda^k(\mathbf{R}_p^n)^*$ be the set of all k-linear alternating maps

$$\varphi: \underbrace{\mathbf{R}_p^n \times \ldots \times \mathbf{R}_p^n}_{k \text{ times}} \to \mathbf{R}$$

(alternating means that φ changes signs with the interchange of two consecutive arguments). With the usual operations, $\Lambda^k(\mathbf{R}_p^n)^*$ is a vector space. Given $\varphi_1, \ldots, \varphi_k \in (\mathbf{R}_p^n)^*$, we can obtain an element $\varphi_1 \wedge \varphi_2 \wedge \ldots \wedge \varphi_k$ of $\Lambda^k(\mathbf{R}_p^n)^*$ by setting

$$(\varphi_1 \wedge \varphi_2 \wedge \ldots \wedge \varphi_k)(v_1, v_2, \ldots, v_k) = \det(\varphi_i(v_j)), \quad i, j = 1, \ldots, k.$$

It follows from the properties of determinants that $\varphi_1 \wedge \varphi_2 \wedge \ldots \wedge \varphi_k$ is in fact k-linear and alternate. In particular $(dx_{i_1})_p \wedge (dx_{i_2})_p \wedge \ldots \wedge (dx_{i_k}) \in \Lambda^k(\mathbf{R}^n_p)^*$, $i_1, i_2, \ldots, i_k = 1, \ldots, n$. We will denote this element by $(dx_{i_1} \wedge dx_{i_2} \wedge \ldots \wedge dx_{i_k})_p$.

Proposition 1. *The set*

$$\{(dx_{i_1} \wedge \ldots \wedge dx_{i_k})_p, \quad i_1 < i_2 < \ldots < i_k, \quad i_j \in \{1, \ldots, n\}\}$$

is a basis for $\Lambda^k(\mathbf{R}^n_p)^*$.

Proof. The elements of the set are linearly independent. For, if

$$\sum_{i_1 < \ldots < i_k} a_{i_1 \ldots i_k} \; dx_{i_1} \wedge \ldots \wedge dx_{i_k} = 0,$$

is applied to

$$(e_{j_1}, \ldots, e_{j_k}), \; j_1 < \ldots < j_k, \; j_\ell \in \{1, \ldots, n\},$$

we obtain (Exercise 2)

$$\sum_{i_1 < \ldots < i_k} a_{i_1 \ldots i_k} \quad dx_{i_1} \wedge \ldots \wedge dx_{i_k} \quad (e_{j_1}, \ldots, e_{j_k}) = a_{j_1 \ldots j_k} = 0.$$

We now show that if $f \in \Lambda^k(\mathbf{R}^n_p)^*$, then f is a linear combination of the form

$$f = \sum_{i_1 < \ldots < i_k} a_{i_1 \ldots i_k} \quad dx_{i_1} \wedge \ldots \wedge dx_{i_k}.$$

For that, set

$$g = \sum_{i_1 < \ldots < i_k} f(e_{i_1}, \ldots, e_{i_k}) \, dx_{i_1} \wedge \ldots \wedge dx_{i_k}.$$

Notice that $g \in \Lambda^k(\mathbf{R}^n_p)^*$ and that

$$g(e_{i_1}, \ldots, e_{i_k}) = f(e_{i_1}, \ldots, e_{i_k}),$$

for all $i_1, \ldots, i_k$. It follows that $f = g$. Setting $f(e_{i_1}, \ldots, e_{i_k}) = a_{i_1 \ldots i_k}$, we obtain the above expression for f. □

Definition 3. An *exterior k-form* in $\mathbf{R}^n$ is a map ω that associates to each $p \in \mathbf{R}^n$ an element $\omega(p) \in \Lambda^k(\mathbf{R}^n_p)^*$; by Proposition 1, ω can be written as

$$\omega(p) = \sum_{i_1 < \ldots < i_k} a_{i_1 \ldots i_k}(p)(dx_{i_1} \wedge \ldots \wedge dx_{i_k})_p, \quad i_j \in \{1, \ldots, n\},$$

where $a_{i_1 \dots i_k}$ are real functions in $\mathbf{R}^n$. When the $a_{i_1 \dots i_k}$ are differentiable functions, ω is called a *differential k-form.*

For notational convenience, we will denote by I the k-upla $(i_1, \dots, i_k)$, $i_1 < \dots < i_k$, $i_j \in \{1, \dots, n\}$, and will use the following notation for ω:

$$\omega = \sum_I a_I dx_I.$$

We also set the convention that a differential 0-form is a differentiable function $f: \mathbf{R}^n \to \mathbf{R}$.

Example 1. In $\mathbf{R}^4$ we have the following types of exterior forms (where a_i, a_{ij}, etc., are real functions in $\mathbf{R}^4$):

0-forms, functions in $\mathbf{R}^4$,
1-forms, $a_1 dx_1 + a_2 dx_2 + a_3 dx_3 + a_4 dx_4$,
2-forms, $a_{12} dx_1 \wedge dx_2 + a_{13} dx_1 \wedge dx_3 + a_{14} dx_1 \wedge dx_4 + a_{23} dx_2 \wedge dx_3 + a_{24} dx_2 \wedge dx_4 + a_{34} dx_3 \wedge dx_4$,
3-forms, $a_{123} dx_1 \wedge dx_2 \wedge dx_3 + a_{124} dx_1 \wedge dx_2 \wedge dx_4 + a_{134} dx_1 \wedge dx_3 \wedge dx_4 + a_{234} dx_2 \wedge dx_3 \wedge dx_4$,
4-forms, $a_{1234} dx_1 \wedge dx_2 \wedge dx_3 \wedge dx_4$.

From now on, we will restrict ourselves to differential k-forms and we will call them simply k-forms.

We are going to define some operations on k-forms in $\mathbf{R}^n$.

First, if ω and φ are two k-forms:

$$\omega = \sum_I a_I dx_I, \qquad \varphi = \sum_I b_I dx_I,$$

we can define their *sum*

$$\omega + \varphi = \sum_I (a_I + b_I) dx_I.$$

Next, if ω is a k-form and φ is an s-form, we can define their *exterior product* $\omega \wedge \varphi$, which is an $(s+k)$-form, as follows.

Definition 4. Let

$$\omega = \sum a_I dx_I, \qquad I = (i_1, \dots, i_k), \qquad i_1 < \dots < i_k,$$

$$\varphi = \sum b_J dx_J, \qquad J = (j_1, \dots, j_s), \qquad j_1 < \dots < j_s.$$

By definition,

$$\omega \wedge \varphi = \sum_{IJ} a_I b_J dx_I \wedge dx_J.$$

Example 2. Let $\omega = x_1dx_1 + x_2dx_2 + x_3dx_3$ be a 1-form in $\mathbf{R}^3$ and $\varphi = x_1dx_1 \wedge dx_2 + dx_1 \wedge dx_3$ be a 2-form in $\mathbf{R}^3$. Then, since $dx_i \wedge dx_i = 0$ and $dx_i \wedge dx_j = -dx_j \wedge dx_i$, $i \neq j$, we obtain

$$\begin{aligned}\omega \wedge \varphi &= x_2dx_2 \wedge dx_1 \wedge dx_3 + x_3x_1dx_3 \wedge dx_1 \wedge dx_2 \\ &= (x_1x_3 - x_2)dx_1 \wedge dx_2 \wedge dx_3.\end{aligned}$$

Remark 1. The definition of exterior product is made in such a way that if $\varphi_1, \dots, \varphi_k$ are 1-forms, then the exterior product $\varphi_1 \wedge \dots \wedge \varphi_k$ agrees with the k-form previously defined by

$$\varphi_1 \wedge \dots \wedge \varphi_k(v_1, \dots, v_k) = \det(\varphi_i(v_j)).$$

This follows immediately from the definition and will be left as an exercise (Exercise 3).

The exterior product of forms in $\mathbf{R}^n$ has the following properties.

Proposition 2. *Let ω be a k-form, φ be an s-form and θ be an r-form. Then:*

a) $(\omega \wedge \varphi) \wedge \theta = \omega \wedge (\varphi \wedge \theta)$,
b) $(\omega \wedge \varphi) = (-1)^{ks}(\varphi \wedge \omega)$,
c) $\omega \wedge (\varphi + \theta) = \omega \wedge \varphi + \omega \wedge \theta$, *if* $r = s$.

Proof. (a) and (c) are straightforward. To prove (b), we write

$$\omega = \sum a_I dx_I, \qquad I = (i_1, \dots, i_k), \qquad i_1 < \dots < i_k,$$

$$\varphi = \sum b_J dx_J, \qquad J = (j_1, \dots, j_s), \qquad j_1 < \dots < j_s.$$

Then

$$\begin{aligned}\omega \wedge \varphi &= \sum_{IJ} a_I b_J dx_{i_1} \wedge \dots \wedge dx_{i_k} \wedge dx_{j_1} \wedge \dots \wedge dx_{j_s} \\ &= \sum_{IJ} b_J a_I (-1) dx_{i_1} \wedge \dots \wedge dx_{i_{k-1}} \wedge dx_{j_1} \wedge dx_{i_k} \wedge \dots \wedge dx_{j_s} \\ &= \sum_{IJ} b_J a_I (-1)^k dx_{j_1} \wedge dx_{i_1} \wedge \dots \wedge dx_{i_k} \wedge dx_{j_2} \wedge \dots \wedge dx_{j_s}.\end{aligned}$$

Since J has s elements, we obtain, by repeating the above argument for each $dx_{j_\ell}, j_\ell \in J$,

$$\begin{aligned}\omega \wedge \varphi &= \sum_{JI} b_J a_I (-1)^{ks} dx_{j_1} \wedge \dots \wedge dx_{j_s} \wedge dx_{i_1} \wedge \dots \wedge dx_{i_k} \\ &= (-1)^{ks} \varphi \wedge \omega.\end{aligned}$$

□

Remark 2. Although $dx_i \wedge dx_i = 0$, it is not true that for any form $\omega \wedge \omega = 0$. For instance, if

$$\omega = x_1 dx_1 \wedge dx_2 + x_2 dx_3 \wedge dx_4,$$

then

$$\omega \wedge \omega = 2x_1 x_2 dx_1 \wedge dx_2 \wedge dx_3 \wedge dx_4.$$

See however Exercise 4.

One of the most important features of differential forms is the way they behave under differentiable maps. Let $f\colon \mathbf{R}^n \to \mathbf{R}^m$ be a differentiable map. Then f induces a map f^* that takes k-forms in $\mathbf{R}^m$ into k-forms in $\mathbf{R}^n$ and is defined as follows. Let ω be a k-form in $\mathbf{R}^m$. By definition, $f^*\omega$ is the k-form in $\mathbf{R}^n$ given by

$$(f^*\omega)(p)(v_1, \dots, v_k) = \omega(f(p))(df_p(v_1), \dots, df_p(v_k)).$$

Here $p \in \mathbf{R}^n$, $v_1, \dots, v_k \in \mathbf{R}^n_p$, and $df_p\colon \mathbf{R}^n_p \to \mathbf{R}^m_{f(p)}$ is the differential of the map f at p. We set the convention that if g is a 0-form,

$$f^*(g) = g \circ f.$$

We are going to show that the operation f^* on forms is equivalent to "substitution of variables". Before that, we need some properties of f^*.

Proposition 3. *Let $f\colon \mathbf{R}^n \to \mathbf{R}^m$ be a differentiable map, ω and φ be k-forms on $\mathbf{R}^m$ and $g\colon \mathbf{R}^m \to \mathbf{R}$ be a 0-form on $\mathbf{R}^m$. Then:*

a) $f^*(\omega + \varphi) = f^*\omega + f^*\varphi$,

b) $f^*(g\omega) = f^*(g) f^*(\omega)$,

c) If $\varphi_1, \dots, \varphi_k$ are 1-forms in $\mathbf{R}^m$, $f^*(\varphi_1 \wedge \dots \wedge \varphi_k) = f^*(\varphi_1) \wedge \dots \wedge f^*(\varphi_k)$.

Proof. The proofs are very simple. Let $p \in \mathbf{R}^n$ and let $v_1, \dots, v_k \in \mathbf{R}^n_p$. Then

(a) $f^*(\omega + \varphi)(p)(v_1, \dots, v_k) = (\omega + \varphi)(f(p))(df_p(v_1), \dots, df_p(v_k)) = (f^*\omega)(p)(v_1, \dots, v_k) + (f^*\varphi)(p)(v_1, \dots, v_k) = (f^*w + f^*\varphi)(p)(v_1, \dots, v_k)$.

(b) $f^*(g\omega)(p)(v_1, \dots, v_k) = (g\omega)(f(p))(df_p(v_1), \dots, df_p(v_k)) = (g \circ f)(p) \cdot f^*\omega(p)(v_1, \dots, v_k) = f^*g(p) \cdot f^*\omega(p)(v_1, \dots, v_k)$.

(c) By omitting the indication of the point p, we obtain

$$\begin{aligned} f^*(\varphi_1 \wedge \dots \wedge \varphi_k)(v_1, \dots, v_k) &= (\varphi_1 \wedge \dots \wedge \varphi_k)(df(v_1), \dots, df(v_k)) \\ &= \det(\varphi_i(df(v_j)) = \det(f^*\varphi_i(v_j)) \\ &= (f^*\varphi_1 \wedge \dots \wedge f^*\varphi_k)(v_1, \dots, v_k). \end{aligned}$$

Remark 3. We will show below (See Proposition 4) that (c) holds not only for 1-forms but for k-forms as well.

We can now present the promised interpretation of f^*. Let $(x_1, \ldots, x_n)$ be coordinates in $\mathbf{R}^n$, $(y_1, \ldots, y_m)$ be coordinates in $\mathbf{R}^m$ and let $f: \mathbf{R}^n \to \mathbf{R}^m$ be written as

$$y_1 = f_1(x_1, \ldots, x_n), \ldots, y_m = f_m(x_1, \ldots, x_n). \qquad (*)$$

Let $\omega = \sum_I a_I dy_I$ be a k-form in $\mathbf{R}^m$. By using the above properties of f^*, we obtain

$$f^*\omega = \sum_I f^*(a_I)(f^*dy_{i_1}) \wedge \ldots \wedge (f^*dy_{i_k}).$$

Since

$$f^*(dy_i)(v) = dy_i(df(v)) = d(y_i \circ f)(v) = df_i(v),$$

we have

$$f^*\omega = \sum_I a_I(f_1(x_1, \ldots, x_n), \ldots, f_m(x_1, \ldots, x_n))df_{i_1} \wedge \ldots \wedge df_{i_k},$$

where f_i and df_i are functions of x_j. Thus to apply f^* to ω is equivalent to "*substitute*" in ω the variables y_i and their differentials by the functions of x_k and dx_k obtained from $(*)$.

Remark 4. In various situations, it is convenient to use differential forms defined only on some open set $U \subset \mathbf{R}^n$ and not on the entire $\mathbf{R}^n$. It is clear that everything done so far extends trivially to this situation.

Example. (Polar coordinates). Let ω be the 1-form in $\mathbf{R}^2 - \{0,0\}$ by

$$\omega = -\frac{y}{x^2+y^2}dx + \frac{x}{x^2+y^2}dy.$$

Let U be the set in the plane (r, θ) given by

$$U = \{r > 0; 0 < \theta < 2\pi\}$$

and let $f: U \to \mathbf{R}^2$ be the map

$$f(r,\theta) = \begin{cases} x = r\cos\theta \\ y = r\,\mathrm{sen}\,\theta \end{cases}$$

Let us compute $f^*\omega$. Since

$$dx = \cos\theta dr - r\,\mathrm{sen}\,\theta d\theta,$$

$$dy = \mathrm{sen}\,\theta dr + r\cos\theta d\theta,$$

we obtain

$$\begin{aligned} f^*\omega &= -\frac{r\,\mathrm{sen}\,\theta}{r^2}(\cos\theta dr - r\,\mathrm{sen}\,\theta d\theta) + \frac{r\cos\theta}{r^2}(\mathrm{sen}\,\theta dr + r\cos\theta d\theta) \\ &= d\theta. \end{aligned}$$

Notice that (*a*) of Proposition 3 states that the addition of differential forms commutes with the "substitution of variables". We will now show that the same holds for the exterior product.

Proposition 4. *Let $f: \mathbf{R}^n \to \mathbf{R}^m$ be a differentiable map. Then*

(a) $f^*(\omega \wedge \varphi) = (f^*\omega) \wedge (f^*\varphi)$, where ω and φ any two forms in $\mathbf{R}^m$.

(b) $(f \circ g)^*\omega = g^*(f^*\omega)$, where $g: \mathbf{R}^p \to \mathbf{R}^n$ is a differentiable map.

Proof. By setting $(y_1, \ldots, y_m) = (f_1(x_1, \ldots, x_n), \ldots, f_m(x_1, \ldots, x_n)) \in \mathbf{R}^m$, $(x_1, \ldots, x_n) \in \mathbf{R}^n$, $\omega = \sum_I a_I dy_I$, $\varphi = \sum_J b_J dy_J$, we obtain

$$\begin{aligned} f^*(\omega \wedge \varphi) &= f^*(\sum_{IJ} a_I b_J dy_I \wedge dy_J) \\ &= \sum_{IJ} a_I(f_1, \ldots, f_m) b_J(f_1, \ldots, f_m) df_I \wedge df_J \\ &= \sum_I a_I(f_1, \ldots, f_m) df_I \wedge \sum_J b_J(f_1, \ldots, f_m) df_J \\ &= f^*\omega \wedge f^*\varphi. \end{aligned}$$

b) $(f \circ g)^*\omega = \sum_I a_I((f \circ g)_1, \ldots, (f \circ g)_m) d(f \circ g)_I$
$= \sum_I a_I(f_1(g_1, \ldots, g_n), \ldots, f_m(g_1, \ldots, g_n)) df_I(dg_1, \ldots, dg_n)$
$= g^*(f^*(\omega))$. □

We are now going to define an operation on differential form that generalizes the differentiation of functions. Let $g: \mathbf{R}^n \to \mathbf{R}$ be a 0-form (i.e., a differentiable function). Then the differential

$$dg = \sum_{i=1}^{n} \frac{\partial g}{\partial x_i} dx_i$$

is a 1-form. We want to generalize this process by defining an operation that takes k-forms into $(k+1)$-forms.

Definition 5. Let $\omega = \sum a_I dx_I$ be a k-form in $\mathbf{R}^n$. The *exterior differential* $d\omega$ of ω is defined by

$$d\omega = \sum_I da_I \wedge dx_I.$$

Example 4. Let $\omega = xyzdx + yzdy + (x+z)dz$ and let us compute $d\omega$:

$$\begin{aligned} d\omega &= d(xyz) \wedge dx + d(yz) \wedge dy + d(x+z) \wedge dz \\ &= (yzdx + xzdy + xydz) \wedge dx + (zdy + ydz) \wedge dy + (dx + dz) \wedge dz \\ &= -xzdx \wedge dy + (1 - xy)dx \wedge dz - ydy \wedge dz. \end{aligned}$$

We now present some properties of exterior differentiation. Item (c) is probably the most important one and item (d) means that the operation d commutes with substitution of variables.

Proposition 5.
a) $d(\omega_1 + \omega_2) = d\omega_1 + d\omega_2$, *where* ω_1 *and* ω_2 *are* k*-forms*
b) $d(\omega \wedge \varphi) = d\omega \wedge \varphi + (-1)^k \omega \wedge d\varphi$, *where* ω *is a* k*-form and* φ *is an* s*-form*
c) $d(d\omega) = d^2\omega = 0$.
d) $d(f^*\omega) = f^*(d\omega)$, *where* ω *is a* k*-form in* $\mathbf{R}^m$ *and* $f: \mathbf{R}^n \to \mathbf{R}^m$ *is a differentiable map.*

Proof.
(a) is straightforward.
(b) Let $\omega = \sum_I a_I dx_I$, $\varphi = \sum_J b_J dx_J$. Then

$$\begin{aligned} &= \sum_{IJ} d(a_I b_J) \wedge dx_I \wedge dx_J \\ &= \sum_{IJ} b_J da_I \wedge dx_I \wedge dx_J + \sum_{IJ} a_I db_J \wedge dx_I \wedge dx_J \\ &= d\omega \wedge \varphi + (-1)^k \sum_{IJ} a_I dx_I \wedge db_J \wedge dx_J \\ &= d\omega \wedge \varphi + (-1)^k \omega \wedge d\varphi. \end{aligned}$$

(c) Let us first assume that ω is a 0-form, i.e., ω is a function $f: \mathbf{R}^n \to \mathbf{R}$ that associates to each $(x_1, \ldots, x_n) \in \mathbf{R}^n$ the value $f(x_1, \ldots, x_n) \in \mathbf{R}$. Then

$$\begin{aligned} d(df) &= d\left(\sum_{j=1}^{n} \frac{\partial f}{\partial x_j} dx_j\right) = \sum_{j=1}^{n} d\left(\frac{\partial f}{\partial x_j}\right) \wedge dx_j \\ &= \sum_{j=1}^{n}\left(\sum_{i=1}^{n} \frac{\partial^2 f}{\partial x_i \partial x_j} dx_i \wedge dx_j\right). \end{aligned}$$

Since $\frac{\partial^2 f}{\partial x_i \partial x_j} = \frac{\partial^2 f}{\partial x_j \partial x_i}$ and $dx_i \wedge dx_j = -dx_j \wedge dx_i$, $i \neq j$, we obtain that

$$d(df) = \sum_{i<j}\left(\frac{\partial^2 f}{\partial x_i \partial x_j} - \frac{\partial^2 f}{\partial x_j \partial x_i}\right) dx_i \wedge dx_j = 0.$$

Now let $w = \sum a_I dx_I$. By (a), we can restrict ourselves to the case $w = a_I dx_I$ with $a_I \neq 0$. By (b), we have that

$$dw = da_I \wedge dx_I + a_I d(dx_I).$$

But $d(dx_I) = d(1) \wedge dx_I = 0$. Therefore,

$$d(dw) = d(da_I \wedge dx_I) = d(da_I) \wedge dx_I + da_I \wedge d(dx_I) = 0,$$

since $d(da_I) = 0$ and $d(dx_I) = 0$, which proves (c).
(d) We will first prove the result for a 0-form. Let $g: \mathbf{R}^m \to \mathbf{R}$ be a differentiable function that associates to each $(y_1, \ldots, y_m) \in \mathbf{R}^m$ the value $g(y_1, \ldots, y_m)$. Then

$$\begin{aligned} f^*(dg) &= f^*\left(\sum_i \frac{\partial g}{\partial y_i} dy_i\right) = \sum_{ij} \frac{\partial g}{\partial y_i}\frac{\partial f_i}{\partial x_j} dx_j \\ &= \sum_j \frac{\partial (g \circ f)}{\partial x_j} dx_j = d(g \circ f) = d(f^* g). \end{aligned}$$

Now, let $\varphi = \sum_I a_I dx_I$ be a k-form. By using the above, and the fact that f^* commutes with the exterior product, we obtain

$$\begin{aligned} d(f^*\varphi) &= d(\sum_I f^*(a_I) f^*(dx_I)) \\ &= \sum_I d(f^*(a_I)) \wedge f^*(dx_I)) = \sum_I f^*(da_I) \wedge f^*(dx_I) \\ &= f^*(\sum_I da_I \wedge dx_I) = f^*(d\varphi) \end{aligned}$$

which proves (d). □

In the exercises that follow we will often use the canonical isomorphism between $\mathbf{R}^n_p$ and its dual $(\mathbf{R}^n_p)^*$ that is established by the natural inner product $\langle\ ,\ \rangle$ of $\mathbf{R}^n$. We recall that if $\{e_i\}$ is the canonical basis of $\mathbf{R}^n$ and $v_1 = \sum a_i e_i$, $v_2 = \sum b_i e_i$ belong to $(\mathbf{R}^n)_p$, then $\langle v_1, v_2 \rangle = \sum a_i b_i$. The above canonical isomorphism takes a vector $v \in \mathbf{R}^n_p$ to an element $\omega \in (\mathbf{R}^n_p)^*$ given by $\omega(u) = \langle v, u \rangle$, for all $u \in \mathbf{R}^n_p$. If we let the point p vary, this establishes a one-to-one correspondence between vector fields in $\mathbf{R}^n$ and exterior 1-forms in $\mathbf{R}^n$; it is easily seen that this correspondence takes differentiable vector fields into differential 1-forms and conversely.

EXERCISES

1) Prove that a bilinear form $\varphi: \mathbf{R}^3 \times \mathbf{R}^3 \to \mathbf{R}$ is alternate if and only if $\varphi(v, v) = 0$, for all $v \in \mathbf{R}^3$.
2) Prove that if $i_1 < i_2 < \ldots < i_k$ and $j_1 < j_2 < \ldots < j_k$, then

$$(dx_{i_1} \wedge \ldots \wedge dx_{i_k})(e_{j_1}, \ldots, e_{j_k}) = \begin{cases} 1, & \text{if } i_1 = j_1, \ldots, i_k = j_k, \\ 0, & \text{otherwise.} \end{cases}$$

Hint: Consider the determinant $\alpha = |dx_{i_\ell}(e_{j_n})|$. If $i_1 > j_1$, then $i_k > \ldots > i_2 > i_1 > j_1$, that is, $dx_{i_\ell}(e_{j_1}) = 0$, for $\ell = 1, \ldots, k$. If $i_1 < j_1$, then $j_k > \ldots > j_2 > j_1 > i_1$, that is, $dx_{i_1}(e_{j_n}) = 0$, for $n = 1, \ldots, k$. In any case, $\alpha = 0$ if $i_1 \neq j_1$. Assume now that $i_1 = j_1$ and $i_2 \neq j_2$. Proceeding as before, we show that $\alpha = 0$. The argument can be easily continued.

3) Prove the statement of Remark 1.
4) Let φ be an exterior k-form, where k is an odd integer. Show that $\varphi \wedge \varphi = 0$.
5) Let φ, ψ and θ the following forms in $\mathbf{R}^3$:

$$\varphi = xdx - ydy,$$
$$\psi = zdx \wedge dy + xdy \wedge dz,$$
$$\theta = zdy.$$

Compute : $\varphi \wedge \psi$, $\theta \wedge \varphi \wedge \psi$, $d\varphi$, $d\psi$, $d\theta$.

6) Let $f: U \subset \mathbf{R}^m \to \mathbf{R}^n$ be a differentiable map. Assume that $m < n$ and let ω be a k-form in $\mathbf{R}^n$, with $k > m$. Show that $f^*\omega = 0$.
7) Let ω be the 2-form in $\mathbf{R}^{2n}$ given by

$$\omega = dx_1 \wedge dx_2 + dx_3 \wedge dx_4 + \ldots + dx_{2n-1} \wedge dx_{2n}.$$

Compute the exterior product of n copies of ω

8) Let $f: \mathbf{R}^n \to \mathbf{R}^n$ be a differentiable map given by

$$f(x_1, \ldots, x_n) = (y_1, \ldots, y_n),$$

and let $\omega = dy_1 \wedge \ldots \wedge dy_n$. Show that

$$f^*\omega = \det(df) dx_1 \wedge \ldots \wedge dx_n.$$

9) Let ν the n-form in $\mathbf{R}^n$ defined by

$$\nu(e_1, \ldots, e_n) = 1,$$

where $\{e_i\}$, $i = 1, \ldots, n$, is the canonical basis of $\mathbf{R}^n$. Show that:

a) If $v_i = \sum a_{ij} e_j$, then

$$\nu(v_1, \ldots, v_n) = \det(a_{ij}) = \mathrm{vol}(v_1, \ldots, v_n).$$

(the form ν is called the *volume element* of $\mathbf{R}^n$)

b) $\nu = dx_1 \wedge \ldots \wedge dx_n$.

10) (*Hodge star operation*). Given a k-form ω in $\mathbf{R}^n$ we will define an $(n-k)$-form $*\omega$ by setting

$$*(dx_{i_1} \wedge \ldots \wedge dx_{i_k}) = (-1)^\sigma (dx_{j_1} \wedge \ldots \wedge dx_{j_{n-k}})$$

and extending it linearly, where $i_1 < \ldots < i_k$, $j_1 < \ldots < j_{n-k}$, $(i_1, \ldots, i_k, j_1, \ldots, j_{n-k})$ is a permutation of $(1, 2, \ldots, n)$, and σ is 0 or 1 according to the permutation is even or odd, respectively. Show that:

a) If $\omega = a_{12} dx_1 \wedge dx_2 + a_{13} dx_1 \wedge dx_3 + a_{23} dx_2 \wedge dx_3$ is a 2-form in $\mathbf{R}^3$, then

$$*\omega = a_{12} dx_3 - a_{13} dx_2 + a_{23} dx_1,$$

b) If $\omega = a_1 dx_1 + a_2 dx_2$ is a 1-form in $\mathbf{R}^2$, then

$$*\omega = a_1 dx_2 - a_2 dx_1.$$

c) $**\omega = (-1)^{k(n-k)}\omega$.

11) (*The divergence*). A differentiable vector field v in $\mathbf{R}^n$ may be considered as a differentiable map $v: \mathbf{R}^n \to \mathbf{R}^n$. We will define a function $\operatorname{div} v: \mathbf{R}^n \to \mathbf{R}$ (to be called the *divergence* of v) as follows:

$$(\operatorname{div} v)(p) = \operatorname{trace}(dv)_p, \quad p \in \mathbf{R}^n,$$

where $(dv)_p: \mathbf{R}^n_p \to \mathbf{R}^n_p$ is the differential of v at p. Show that:

a) If $v = \sum a_i e_i$, where $\{e_i\}$ is the canonical basis of $\mathbf{R}^n$, then

$$\operatorname{div} v = \sum \frac{\partial a_i}{\partial x_i}.$$

b) If ω denotes the differential 1-form obtained from v by the canonical isomorphism induced by the inner product $\langle\, ,\, \rangle$ and ν is the volume element of $\mathbf{R}^n$ (Exercise 9), the divergence can be obtained as follows:

$$v \longrightarrow \omega \longrightarrow *\omega \longrightarrow d(*\omega) = (\operatorname{div} v)\nu,$$

where we have used the star operation introduced in Exercise 10.

12) (*The gradient*). Given a differentiable function $f: \mathbf{R}^n \to \mathbf{R}$, define a vector field $\operatorname{grad} f$ in $\mathbf{R}^n$ (the *gradient* of f) by

$$\langle \operatorname{grad} f(p), u \rangle = df_p(u), \text{ for all } p \in \mathbf{R}^n \text{ and all } u \in \mathbf{R}^n_p.$$

Notice that $\operatorname{grad} f$ is the vector field corresponding to the 1-form df in the canonical isomorphism. Show that:

a) In the canonical basis $\{e_i\}$ of $\mathbf{R}^n$:

$$\operatorname{grad} f = \sum \frac{\partial f}{\partial x_i} e_i.$$

b) If $p \in \mathbf{R}^n$ is such that $\operatorname{grad} f(p) \neq 0$, then $\operatorname{grad} f(p)$ is perpendicular to the "level surface" $\{q \in \mathbf{R}^n; f(q) = f(p)\}$.

c) The linear map $df_p: \mathbf{R}^n_p \to \mathbf{R}$ restricted to unit sphere with center at p reaches its maximum for $v = \operatorname{grad} f / |\operatorname{grad} f|$.

13) (*The Laplacian*). Given a differentiable function $f: \mathbf{R}^n \to \mathbf{R}$, we will define the *Laplacian* $\Delta f: \mathbf{R}^n \to \mathbf{R}$ by

$$\Delta f = \operatorname{div}(\operatorname{grad} f).$$

Show that:

a) $\Delta f = \sum \frac{\partial^2 f}{\partial x_i^2}$,

b) $\Delta(fg) = f\Delta g + g\Delta f + 2\langle \operatorname{grad} f, \operatorname{grad} g\rangle$,

c) $d * (df) = (\Delta f)\nu$, where ν is the volume element of $\mathbf{R}^n$.

14) (*The rotational*). Let v be a differentiable vector field in $\mathbf{R}^n$. The rotational rot v is the $(n-2)$-form defined by:

$$v \longmapsto \omega \longrightarrow d\omega \longrightarrow *(d\omega) = \text{ rot } v,$$

where $v \mapsto \omega$ is the correspondence between 1-forms and vector fields induced by the inner product.

a) Prove that $\operatorname{rot}(\operatorname{grad} f) = 0$.

b) In the particular case when $n = 3$, the 1-form rot v corresponds to a vector field which is also denoted by rot v. Show that, for $n = 3$:

b1)

$$\begin{aligned}\operatorname{rot}\Big(\sum_{i=1}^{3} a_i e_i\Big) &= \left(\frac{\partial a_3}{\partial x_2} - \frac{\partial a_2}{\partial x_3}\right) e_1 \\ &+ \left(\frac{\partial a_1}{\partial x_3} - \frac{\partial a_3}{\partial x_1}\right) e_2 \\ &+ \left(\frac{\partial a_2}{\partial x_1} - \frac{\partial a_1}{\partial x_2}\right) e_3.\end{aligned}$$

b2) $\operatorname{div}(\operatorname{rot} v) = 0$.

15) (*A geometric definition of the $*$ operation*). An element $\varphi \in \Lambda^k(\mathbf{R}^n_p)^*$ is called decomposable if $\varphi = \varphi_1 \wedge \ldots \wedge \varphi_k$, where φ_i, $i = 1, \ldots, k$, are linearly independent elements of $\Lambda^1(\mathbf{R}^n_p)^* \simeq (\mathbf{R}^n_p)^*$. Prove that:

a) If $\varphi_i = \sum_j a_{ij}\beta_j$, $i = 1, \ldots, k$, $\beta_j \in (\mathbf{R}^n_p)^*$, and $\det(a_{ij}) = 1$, then $\varphi_1 \wedge \ldots \wedge \varphi_k = \beta_1 \wedge \ldots \wedge \beta_k$; thus, a decomposable element may have more than one representation.

b) If $\varphi_1 \wedge \ldots \wedge \varphi_k = \beta_1 \wedge \ldots \wedge \beta_k = \varphi$ are two representations of φ, then $\varphi_i = \sum_j a_{ij}\beta_j$, with $\det(a_{ij}) = 1$.

Hint: Extend the β_j into a basis $\beta_1, \ldots, \beta_k, \beta_{k+1}, \ldots \beta_n$ of $(\mathbf{R}^n_p)^*$ and write

$$\varphi_i = \sum_j a_{ij}\beta_j + \sum_\eta b_{i\eta}\beta_\eta, \qquad \eta = k+1, \ldots, n.$$

Notice that $\beta_1 \wedge \cdots \wedge \beta_k \wedge \varphi_i = \varphi_1 \wedge \cdots \wedge \varphi_k \wedge \varphi_i = 0$. This implies that

$$\sum_\eta b_{i\eta}\beta_1 \wedge \ldots \wedge \beta_k \wedge \beta_\eta = 0,$$

and since $\beta_1 \wedge \ldots \wedge \beta_k \wedge \beta_\eta$ are linearly independent, $b_{i\eta} = 0$.

c) If $\varphi = \varphi_1 \wedge \ldots \wedge \varphi_k$ is decomposable, the vectors $v_1, \ldots, v_k \in \mathbf{R}^n_p$, where $v_i \leftrightarrow \varphi_i$ is the correspondence induced by the inner product $\langle\ ,\ \rangle$ of $\mathbf{R}^n$, are linearly independent, and the subspace of $\mathbf{R}^n$ generated by them does not depend on the representation of φ. This will be called the *subspace* of φ.

d) If $\varphi = \varphi_1 \wedge \ldots \wedge \varphi_k$ is decomposable, the k-volume of the solid generated by $v_1, \ldots, v_k$ does not depend on the representation of φ. This will be called the *volume* of φ.

e) If $\varphi = \varphi_1 \wedge \ldots \wedge \varphi_k$ is decomposable, define $*\varphi$ as an element of $\wedge^{n-k}(\mathbf{R}^n_p)^*$ with the following properties:
i) the subspace of $*\varphi$ is perpendicular to the subspace of φ.
ii) the volume of $*\varphi$ is equal to the volume of φ.
iii) $\varphi \wedge *\varphi$ is positive, i.e., its value in a positive basis of $\mathbf{R}^n_p$ is positive.
Prove: That $*\varphi$ is well defined.
Hint: Let $v_i \leftrightarrow \varphi_i$, $i = 1, \ldots, k$, as in (c) and let W be the subspace of $\mathbf{R}^n_p$ generated by the v_i's. In $W^\perp$ consider an orthonormal basis $e_{k+1}, \ldots, e_n$ so that the basis $v_1, \ldots, v_r, e_{k+1}, \ldots, e_n$ of $\mathbf{R}^n_p$ is positive. Define $\varphi_j \in (\mathbf{R}^n_p)^*, j = k+1, \ldots, n$, by the correspondence $\varphi_j \leftrightarrow e_j$ mentioned in (c), and let $\lambda > 0$ be the volume of W. Check that $\lambda\varphi_{k+1}, \wedge \ldots \wedge \varphi_n$ satisfies (i), (ii) and (iii).

f) Let v_1, v_2 be two vectors in $\mathbf{R}^3$ and let $\varphi_1 \leftrightarrow v_1$, $\varphi_2 \leftrightarrow v_2$ be the one-forms given by the correspondence in (c). Define the *vector product* $v_1 \times v_2 \leftrightarrow *(\varphi_1 \wedge \varphi_2)$ and describe geometrically the vector $v_1 \times v_2$.

g) A k-form w in $\mathbf{R}^n$ is decomposable if $w(p)$ is decomposable for each $p \in \mathbf{R}^n$. Every k-form in $\mathbf{R}^n$ is a linear combination of decomposable k-forms of the type $dx_{i_1} \wedge \ldots \wedge dx_{i_k}$. Show that with the above definition, $*(dx_{i_1} \wedge \ldots \wedge dx_{i_k})$ gives the same expression as in Exercise 10.

16) (*Poincaré's lemma for 1-forms*). Let

$$\omega = a(x,y,z)dx + b(x,y,z)dy + c(x,y,z)dz$$

be a differentiable 1-form in $\mathbf{R}^3$ such that $d\omega = 0$. Define $f\colon \mathbf{R}^3 \to \mathbf{R}$ by

$$f(x,y,z) = \int_0^1 (a(tx,ty,tz)x + b(tx,ty,tz)y + c(tx,ty,tz)z)\,dt.$$

Show that $df = \omega$ (This means that the condition $d\omega = 0$, which is satisfied when $\omega = df$, is in $\mathbf{R}^3$ also sufficient for the existence of an f such that $\omega = df$. We have used $\mathbf{R}^3$ for notational convenience but the result holds for $\mathbf{R}^n$. Also, if ω is only defined in an open set $U \subset \mathbf{R}^n$, the result still holds in a neighborhood of each $p \in U$).
Hint: Notice that $d\omega = 0$ implies that

$$\frac{\partial b}{\partial x} = \frac{\partial a}{\partial y}, \quad \frac{\partial c}{\partial x} = \frac{\partial a}{\partial z}, \quad \frac{\partial b}{\partial z} = \frac{\partial c}{\partial y}$$

and use the identity

$$a(x,y,z) = \int_0^1 \frac{d}{dt}(a(tx,ty,tz)t)dt$$
$$= \int_0^1 a(tx,ty,tz)dt + \int_0^1 t(a_1x + a_2y + a_3z)dt,$$

where a_1, a_2 and a_3 denote the partial derivatives of $a(tx,ty,tz)$ relative to the first, second and third argument, respectively.

17) We say that a differentiable vector field v defined in an open set $U \subset \mathbf{R}^n$ *derives locally from a potential* if for each $p \in U$ then exists a neighborhood $V \ni p$, $V \subset U$, and a differentiable function $g: V \to \mathbf{R}$ (to be called the potential) such that $v = \operatorname{grad} g$.
   a) Let v be as above and let ω be the 1-form corresponding to v, i.e., $\omega(u) = \langle v, u\rangle$, for all $u \in \mathbf{R}^n$. Show that v derives locally from a potential if and only if $d\omega = 0$.
   Hint: Use the local form of Exercise 16.
   b) Show that v derives locally from a potential if and only if $\operatorname{rot} v = 0$.
   c) Let v be the vector field (electric attraction):

$$v(p) = -\frac{1}{(x^2+y^2+z^2)^{3/2}}(x,y,z), \qquad p \in \mathbf{R}^3 - \{0,0,0\}$$

   Show that v derives locally from a potential g:

$$g = \frac{1}{(x^2+y^2+z^2)^{1/2}} + \text{const.}$$

   and that $\Delta g = 0$.

18) A function $g : \mathbf{R}^3 \to \mathbf{R}$ is said to be homogeneous of degree k if $g(tx,ty,tz) = t^k g(x,y,z)$, $t > 0$, $(x,y,z) \in \mathbf{R}^3$. Prove that:
   a) If g is differentiable and homogeneous of degree k, then (*Euler's relation*)

$$xg_x + yg_y + zg_z = kg.$$

   Hint: Differentiate $g(tx,ty,tz) = t^k g(x,y,z)$ in t and set $t = 1$.
   b) If the differential form

$$\omega = adx + bdy + cdz$$

   is such that a, b and c are homogeneous of degree k and $d\omega = 0$, then $\omega = df$, where

$$f = \frac{xa + yb + zc}{k+1}$$

   Hint: Notice that $d\omega = 0$ implies that

$$\frac{\partial b}{\partial x} = \frac{\partial a}{\partial y}, \quad \frac{\partial c}{\partial x} = \frac{\partial a}{\partial z}, \quad \frac{\partial b}{\partial z} = \frac{\partial c}{\partial y},$$

   and apply Euler's relation

c) If the differential form

$$\sigma = a dy \wedge dz + b dz \wedge dx + c dx \wedge dy$$

is such that a, b and c are homogeneous of degree k and $d\sigma = 0$, then $\sigma = d\gamma$, where

$$\gamma = \frac{(zb - yc)dx + (xc - za)dy + (ya - xb)dz}{k+2}.$$

2. Line Integrals*

Differential forms are to be integrated. We will do that soon (Chapter 4) after some preliminaries on the natural "habitat" of differential forms (Chapter 3). However, the special case of integration of forms of degree one along curves (the so called line integrals) is so simple that it can be treated independently of the general theory. We will do that in this chapter.

Although we restrict ourselves to curves in $\mathbf{R}^n$, proofs are organized in such a way that will be valid in a more general setting to be considered later.

Let $\omega = \sum a_i dx_i$ be a differential form of degree one defined in an open set $U \subset \mathbf{R}^n$ and let $c: [a, b] \to U$ be a piecewise differentiable curve in U; we recall that c is piecewise differentiable if c is continuous and there exists a partition $a = t_0, t_1, \ldots, t_k, t_{k+1} = b$ of $[a, b]$ such that the restriction $c|[t_j, t_{j+1}] = c_j$ is differentiable, $j = 0, 1, \ldots, k$. Notice that in each interval $[t_j, t_{j+1}]$, $c_j^*\omega$ is a form in $\mathbf{R}$ given by

$$c_j^*\omega = \sum_i a_i(x_1(t), \ldots, x_n(t)) \frac{dx_i}{dt} dt.$$

where $c(t) = (x_1(t), \ldots, x_n(t))$. Define

$$\int_{c(t)} \omega = \sum_j \int_{t_j}^{t_j+1} c_j^*\omega = \int_a^b \left(\sum_i a_i(t) \frac{dx_i}{dt} \right) dt.$$

A *change of parametrization* of $c: [a, b] \to U$ is a differentiable homeomorphism $\varphi: [c, d] \to [a, b]$. We say that φ *preserves orientation* if φ is increasing; otherwise, it *reverses orientation*. If $t = \varphi(\tau)$ and φ is increasing, we obtain, using the formula of change of variables for integrals,

$$\begin{aligned} \int_{c(t)} \omega &= \int_a^b \left(\sum a_i(t) \frac{dx_i}{dt} \right) dt = \int_a^b \left(\sum a_i(\varphi(\tau)) \frac{dx_i}{d\tau} \frac{d\tau}{dt} \right) dt \\ &= \int_c^d \sum a_i(\tau) \frac{dx_i}{d\tau} d\tau = \int_{c(\tau)} \omega, \end{aligned}$$

* This chapter is not used in the rest of the book and can be omitted on a first reading.

which shows that $\int_c \omega$ is invariant by a change of parametrization that preserves orientation; similarly, if φ changes orientation, $\int_c \omega$ changes sign.

We will denote by c the trace of $c(t)$ with a given orientation and by $-c$ the same curve with the opposite orientation. Then $\int_c \omega$ is well defined and

$$\int_{-c} \omega = -\int_c \omega.$$

We will say that ω is *closed* if $d\omega = 0$, and that ω is *exact* in $V \subset U$ if there exist a differentiable function $f: V \to \mathbf{R}$ such that $\omega = df$ in V. Notice that if ω is exact in V, and $c: [a,b] \to V$ is a curve,

$$\int_c \omega = \int_c df = \int_a^b c^*(df) = f(c(b)) - f(c(a)),$$

that is, $\int_c \omega$ depends only on the end points of c. It also follows that if ω is exact in V and c is a closed curve in V, then $\int_c \omega = 0$.

Actually, these three properties are equivalent. From now on, $\omega = \sum a_i dx_i$ is a differential form defined in an open set $U \subset \mathbf{R}^n$.

Proposition 1. *The following are equivalent:*

1) ω *is exact in a connected open set* $V \subset U$.

2) $\int_c \omega$ *depends only on the end points of* c *for all* $c \subset V$.

3) $\int_c \omega = 0$, *for all closed curves* $c \subset V$.

Proof. We have already shown that (1) $\Rightarrow$ (2) $\Rightarrow$ (3). That (3) $\Rightarrow$ (2) is immediate. It remains to show that (2) $\Rightarrow$ (1).

Let us assume (2) and fix a point $p \in V$. For each $x \in V$, let c be a piecewise differentiable curve joining p to x. Define $f: V \to \mathbf{R}$ by $f(x) = \int_c \omega$. By (2), f is well defined. We claim that $df = \omega$, which will conclude the proof.

Since $df = \sum_i \frac{\partial f}{\partial x_i} dx_i$, we must show that $\frac{\partial f}{\partial x_i}(x) = a_i(x)$, $i = 1, \ldots, n$. Let $e_i = (0, \ldots, 0, 1, 0, \ldots, 0)$, where 1 is in i-th place, and consider the curve

$$c_i: t \to x + te_i, \quad t \in (-\varepsilon, \varepsilon),$$

that joins x to $x + te_i$ and is contained in V for t small. Then

$$\begin{aligned}\frac{\partial f}{\partial x_i}(x) &= \lim_{t\to 0}\frac{1}{t}\{f(x+te_i)-f(x)\}\\ &= \lim_{t\to 0}\frac{1}{t}\left\{\int_{c+c_i}\omega - \int_c \omega\right\} = \lim_{t\to 0}\frac{1}{t}\int_{c_i}\omega\\ &= \lim_{t\to 0}\frac{1}{t}\int_0^t a_i(s)ds = a_i(0) = a_i(x).\end{aligned}$$

□

Example 1. Consider the form

$$\omega_0 = -\frac{y}{x^2+y^2}dx + \frac{x}{x^2+y^2}dy$$

defined in $U = \mathbf{R}^2 - \{(0,0)\}$. A direct computation shows that $d\omega_0 = 0$. Rather than doing this computation, we prefer a geometric approach. Choose a half-line L issuing from the origin $0 = \{(0,0)\}$ and consider polar coordinates (ρ, θ) in $\mathbf{R}^2 - L$. Since $x = \rho\cos(\theta+\theta_0)$, $y = \rho\sin(\theta+\theta_0)$, we obtain that $\omega_0 = d\theta$ in $\mathbf{R}^2 - L$; here θ_0 is the angle from Ox to L. Since L is arbitrary, $d\omega_0 = d^2\theta = 0$, hence ω_0 is closed. In addition, this shows that the form ω_0 is locally (that is, in a neighborhood V of each point of U) the differential of a function θ that measures the positive angle of $\overrightarrow{O_p}$ with L, $p \in V$. We call θ an *angle function* and ω_0 *the element of angle* (with respect to the origin 0).

Although the form ω_0 is closed and locally exact, it is not exact in $\mathbf{R}^2 - \{0\}$. To see this, consider the unit circle $c(t) = (\cos t, \sin t)$, $t \in [0, 2\pi]$, around the origin. Then $c\colon [0,2\pi] \to U$, and

$$\int_{c(t)}\omega_0 = \int_0^{2\pi} c^*\omega_0 = \int_0^{2\pi} dt = 2\pi.$$

It follows by (3) of Proposition 1 that ω_0 is not exact in $\mathbf{R}^2 - \{0\}$. This means that it is not possible to patch together the various angle functions locally defined to make an angle function globally defined in $\mathbf{R}^2 - \{0\}$.

Remark 1. (*The winding number*) Although ω_0 is not exact in $\mathbf{R}^2 - \{0\}$, it determines, along a given curve $\gamma\colon [0,1] \to \mathbf{R}^2 - \{0\}$, a well defined "angle function" $\varphi(t)$, $t \in [0,1]$, given by

$$\varphi(t) = \int_0^1 \frac{xy' - yx'}{x^2+y^2}dt + \varphi_0, \quad \gamma(t) = (x(t), y(t)).$$

Setting

$$a(t) = \frac{x}{\sqrt{x^2+y^2}}(t), \quad b(t) = \frac{y}{\sqrt{x^2+y^2}}(t),$$

it is easily checked that $\gamma^*\omega_0 = (ab' - ba')dt$, where $a^2 + b^2 = 1$. It can be shown that if φ_0 is such that

$$\cos\varphi_0 = a(0), \quad \sin\varphi_0 = b(0),$$

then $\cos\varphi(t) = a(t)$, $\sin\varphi(t) = b(t)$ (for details, see M. do Carmo **[dC]**, Lemma 1, p. 250, or Lemma 5 of Chapter 5 below). Thus $\varphi(t)$ is a continuous determination of the angle that $\gamma(t)$ makes with $\gamma(t_0)$. If γ is closed (i.e., $\gamma(1) = \gamma(0)$), then $\varphi(1) \neq \varphi(0)$, but since $\cos\varphi(1) = \cos\varphi(0)$ and $\sin\varphi(1) = \sin\varphi(0)$, we obtain that $\varphi(1) - \varphi(0)$ is an integral multiple of 2π. This integer is usually called the *winding number* of γ around 0.

The fact that the closed form ω_0 in Example 1 is locally exact is a general fact. More precisely.

Theorem 1. *(Poincaré's Lemma for 1-forms) Let $\omega = \sum a_i dx_i$ be defined in an open set $U \subset \mathbf{R}^n$. Then $d\omega = 0$ if and only if for each $p \in U$ there is a neighborhood $V \subset U$ of p and a differentiable function $f: V \to \mathbf{R}$ with $df = \omega$ (i.e., ω is locally exact).*

Proof. If ω is locally exact, clearly $d\omega = 0$. Let us now assume that $d\omega = 0$. For simplicity of notation, let us restrict ourselves to the case where $\omega = a dx + b dy + c dz$ is defined in $U \subset \mathbf{R}^3$. For each $p \in U$, let $B(p)$ be a ball of center $p = (x_0, y_0, z_0)$ contained in U. For each $q \in B(p), q = (x, y, z)$, let $\beta(t) = p + t(q - p)$, $t \in [0, 1]$, be the line that joins p to q. Since $B(p)$ is a ball, $\beta(t) \subset B(p)$. Define

$$f(q) = \int_{\beta(t)} \omega = \int_0^1 \{a(\beta(t))(x - x_0) + b(\beta(t))(y - y_0) + c(\beta(t))(z - z_0)\} dt$$

We want to show that $df = \omega$, that is,

$$\frac{\partial f}{\partial x}(q) = a(q), \quad \frac{\partial f}{\partial y}(q) = b(q), \quad \frac{\partial f}{\partial z}(q) = c(q).$$

To see this, notice that the condition that $d\omega = 0$ is equivalent to:

$$\frac{\partial a}{\partial y} = \frac{\partial b}{\partial x}, \quad \frac{\partial a}{\partial z} = \frac{\partial c}{\partial x}, \quad \frac{\partial b}{\partial z} = \frac{\partial c}{\partial y}.$$

Let us consider first the case $\frac{\partial f}{\partial x} = a$. Differentiating f and using the first two identities above, we obtain

$$\begin{aligned}
\frac{\partial f}{\partial x}(q) &= \int_0^1 \left\{ \frac{\partial a}{\partial x} t(x - x_0) + a + \frac{\partial b}{\partial x} t(y - y_0) + \frac{\partial c}{\partial x} t(z - z_0) \right\} dt = \\
&= \int_0^1 \left\{ \left(\frac{\partial a}{\partial x}(x - x_0) + \frac{\partial a}{\partial y}(y - y_0) + \frac{\partial a}{\partial z}(z - z_0) \right) t + a \right\} dt = \\
&= \int_0^1 \left\{ \left(\frac{d}{dt}(a(\beta(t))) \right) t + a \right\} dt = \\
&= \int_0^1 \frac{d}{dt}(a(\beta(t))t) dt = a(\beta(1)) = a(q).
\end{aligned}$$

In a similar way, we can prove that

$$\frac{\partial f}{\partial y}(q) = b(q), \quad \frac{\partial f}{\partial z}(q) = c(q).$$

This completes the proof of Theorem 1. □

One of the interesting applications of Theorem 1 is to extend the definition of the integral of a closed form to curves that are merely continuous. To do that, observe first that if ω is defined in $U \subset \mathbf{R}^n$ is closed and $c: [0,1] \to U$ is a differentiable curve, we can choose a partition

$$0 = t_0 < t_1 < \cdots < t_k < t_{k+1} = 1$$

of [0,1] in such a way that the restriction $c|[t_i, t_{i+1}] = c_i$, $i = 0, 1, \ldots, k$, is contained in a ball B_i where ω is exact, i.e., there exists a function $f_i: B_i \to \mathbf{R}$ with $df_i = \omega$. Then

$$\int_c \omega = \sum_i \int_{c_i} \omega = \sum_i [f_i(t_{i+1}) - f_i(t_i)]. \tag{1}$$

Now, if $c: [0,1] \to U$ is only continuous, such a partition still exists and we define $\displaystyle\int_c \omega$ by (1).

We must show that this is independent of the chosen subdivision. This is a standard argument and we will present it here for completeness.

Given a partition P, a refinement of P is a new partition obtained from P by adding new points to it. If we add a point $t' \in (t_i, t_{i+1})$, we have that $c(t) \in B_i$. Since

$$[f_i(t_{i+1}) - f_i(t')] + [f_i(t') - f_i(t_i)] = [f_i(t_{i+1}) - f_i(t_i)],$$

the value of the integral for this new partition does not change. It follows that the integral remains the same under refinements of a given partition. Now, given two distinct partitions, we can form a third partition, by adding to the first all the points of the second. This is a common refinement, the integral of which must be equal to the integrals of the first and of the second partitions, which are therefore equal. This shows the required independence.

Theorem 1 also raises the following question. We know that a closed form is locally exact. When is it globally exact? We know that we need some restriction on the domain of the closed form, since by Example 1, the form element of angle is closed in $\mathbf{R}^2 - \{0\}$ but it is not exact there.

The answer to this question will depend on how the integral of a closed form along a continuous curve varies when we deform the curve continuously. To make this notion precise, we introduce the notion of homotopy.

Definition 1. Two continuous curves $c_0, c_1 \colon [a,b] \to U \subset \mathbf{R}^n$ with the same end points $c_0(a) = c_1(a)$ and $c_0(b) = c_1(b)$ are *homotopic* if there exists a continuous map

$$H \colon [a,b] \times [0,1] \to U, \quad (s,t) \in [a,b] \times [0,1],$$

such that:

$$H(s,0) = c_0(s), \quad H(s,1) = c_1(s), \tag{2}$$

$$H(a,t) = c_0(a) = c_1(a), \quad H(b,t) = c_0(b) = c_1(b). \tag{3}$$

Thus the homotopy $H(s,t) = H_t(s)$ is a continuous family of curves parametrized by $t \in [0,1]$ that deforms the curve $H_0(s) = c_0(s)$ into the curve $H_1(s) = c_1(s)$ (condition (2)), by keeping fixed the end points $H_t(a)$ and $H_t(b)$ (condition (3)).

Sometimes it is convenient to drop condition (3) in the definition of homotopy and to let the end points vary; in this case, we say that H is a *free homotopy* between c_0 and c_1.

It turns out that line integrals of closed forms are invariant under homotopies. More precisely.

Theorem 2. *Let ω be a closed 1-form defined in an open set $U \subset \mathbf{R}^n$. Let $c_0, c_1 \colon [a,b] \to U$ be two continuous homotopic curves in U. Then*

$$\int_{c_0} \omega = \int_{c_1} \omega. \tag{4}$$

Proof. Since $d\omega = 0$, ω is locally exact. Let H be a homotopy between c_0 and c_1, and let $\{B_i\}$ be a covering of $H([a,b] \times [0,1]) \subset U$ by balls B_i where ω is exact. Since $[a,b] \times [0,1] = R$ is compact, the covering of R by $\{H^{-1}(B_i)\} = W_i$ has a Lebesgue number d (that is, each subset of R with diameter $< d$ is contained in some W_i). Divide R into subrectangles R_{jk} by the lines $s =$ const., $t =$ const., in such a way that the diameter of each R_{jk} is smaller than d. By local exactness, $\int_{\partial R_{jk}} \omega = 0$. Therefore, if R_{jk} has sides $\alpha_{jk}, \beta_{j,k+1}, \alpha_{j+1,k}, \beta_{jk}$, with the orientations given by increasing s and increasing t, we obtain (Fig. 2.1).

$$0 = \sum_{jk} \int_{\partial R_{jk}} \omega = \sum_{jk} \left\{ \int_{\alpha_{jk}} \omega + \int_{\beta_{j,k+1}} \omega - \int_{\alpha_{j+1,k}} - \int_{\beta_{jk}} \omega \right\}.$$

But the sides of R_{jk} that are interior to R appear twice with opposite orientations in the above sum. So the corresponding integrals cancel out, and we have

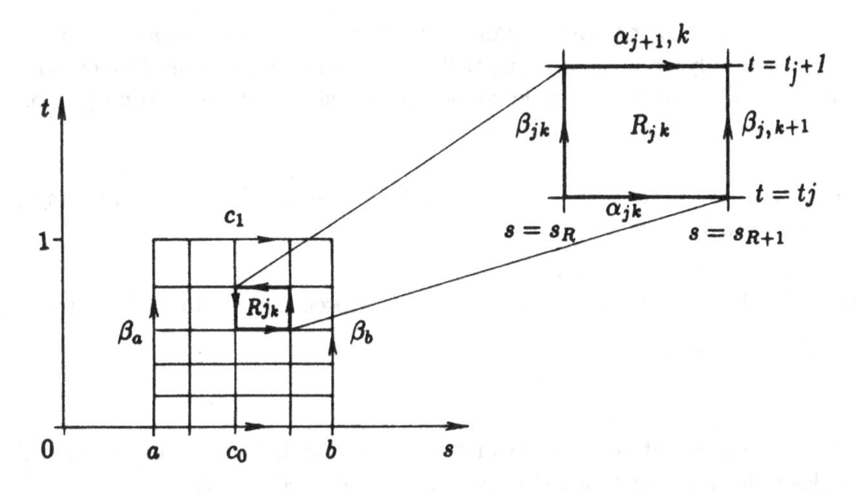

Fig. 2.1

$$0 = \int_{c_0} \omega + \int_{\beta_b} \omega - \int_{c_1} \omega - \int_{\beta_a} \omega, \qquad ((5))$$

where β_a is the curve $H(a,t)$ and β_b is the curve $H(b,t)$. Since these curves reduce to points, the corresponding integrals vanish, and we are left with

$$\int_{c_0} \omega = \int_{c_1} \omega,$$

as we wished. □

We should observe that if c_0 and c_1 are closed curves (i.e., $c_0(a) = c_0(b)$ and $c_1(a) = c_1(b)$) that are freely homotopic, then, although the curves β_a and β_b do not reduce to points, they are equal. Therefore, by (5), we still have the equality in (4). For future reference, we will sum up the above discussion in a proposition.

Proposition 2. *Let ω be a closed 1-form defined in U, and let c_0 and c_1 be two closed curves freely homotopic in U. Then $\int_{c_0} \omega = \int_{c_1} \omega$; in particular, if c_0 is freely homotopic to a point, $\int_{c_0} \omega = 0$.*

We will say that a connected open set $U \subset \mathbf{R}^n$ is *simply-connected* if every continuous closed curve in U is freely homotopic to a point in U. This is a topological property of a domain $U \subset \mathbf{R}^n$. For example, the space $\mathbf{R}^n$ itself, the balls of $\mathbf{R}^n$ and its homeomorphic images are all simply connected domains in $\mathbf{R}^n$. On the other hand, as we will show in a moment, $\mathbf{R}^2 - \{0\}$ is not simply–connected. This will follow from Proposition 3 below and Example 1. The notion of simple connectivity yields a sufficient condition on a domain for a closed form to be exact.

Proposition 3. *Let ω be a closed form defined in a simply-connected domain. Then ω is exact.*

Proof. By Proposition 2, $\int_c \omega = 0$ for every closed curve c in U. By Proposition 1, ω is exact. □

A further application of the invariance under free homotopy of the integral of a closed form along a closed curve can be obtained as follows.

Let $F: U \subset \mathbf{R}^2 \to \mathbf{R}^2$ be a differentiable map. We say that $p \in U$ is a *zero* of F if $F(p) = 0$. If there exists a neighborhood V of p such that V contains no zero of F other than p, then p is called an *isolated zero*. If the differential $dF(p)$ of F at p is nonsingular, we say that p is a *simple zero* of F. By the inverse function theorem, F is one-to-one in a neighborhood of a simple zero, hence a simple zero is isolated.

Now let $F(x,y) = (f(x,y), g(x,y))$ and let $D \subset U$ be a closed disk such that the boundary $\partial D = C$ contains no zeroes of F. Consider the differential form

$$\theta = \frac{f dg - g df}{f^2 + g^2}$$

defined at the points of U where $f^2 + g^2 \neq 0$. We will call

$$\frac{1}{2\pi} \int_{\partial D} \theta = n(F; D)$$

the *index of F in D*. We claim that $n(F; D)$ is an integer.

To see that, let $u = f(x,y)$, $v = g(x,y)$, and let $\omega_0 = \frac{udv - vdu}{u^2+v^2}$ be the element of angle in the plane (u,v) relative to the origin (0,0). Then $\theta = F^*\omega_0$ and

$$n(F; D) = \frac{1}{2\pi} \int_C \theta = \frac{1}{2\pi} \int_C F^*\omega_0 = \frac{1}{2\pi} \int_{F \circ C} \omega_0,$$

that is, $n(F; D)$ is the winding number (see Remark 1) of the closed curve $F \circ C$ around the origin in the plane (u,v). This shows that $n(F; D)$ is an integer.

A consequence of the invariance under free homotopy of $\int_C \theta$ is the following existence theorem for solutions of the equation $F = 0$.

Proposition 4. *If $n(F; D) \neq 0$, there exists some point $q \in D$ such that $F(q) = 0$.*

Proof. Assume that no such q exists, and let p be the center of the disk D. Then the map $H: [0, 2\pi] \times [0, 1] \to \mathbf{R}^2 - \{0\}$ given by

$$H(s, t) = F((1 - t)C(s) + tp)$$

is a free homotopy between the curve $F \circ C$ and the constant curve $F(p)$. Therefore,

$$n(F; D) = \frac{1}{2\pi} \int_{\partial D} \theta = \frac{1}{2\pi} \int_{F \circ C} \omega_0 = 0$$

which is a contradiction. □

It is clear that if we take another disk $D_1 \supset D$ such that F has no zeroes in $D_1 - \text{int } D$ then $n(F; D_1) = n(F; D)$. This follows because there is an obvious free homotopy between $F(\partial D)$ and $F(\partial D_1)$.

Furthermore, if $F_1, F_2: U \subset \mathbf{R}^2 \to \mathbf{R}^2$ are two differentiable maps with no zeroes in $C = \partial D$, $D \subset U$, and there exists a continuous map $H: C \times [0, 1] \to \mathbf{R}^2 - \{0\}$ with $H(q, 0) = F_1(q)$, $H(q, 1) = F_2(q)$, $H(q, t) \neq 0$, for all $q \in C$, $t \in [0, 1]$, then

$$n(F_1; D) = n(F_2; D).$$

This follows because the curves $F_1 \circ C$, $F_2 \circ C$ are freely homotopic by

$$K: [0, 2\pi] \times [0, 1] \to \mathbf{R}^2 - \{0\}, \quad K(s, t) = H(C(s), t).$$

A most interesting application of the index is the formula below that is attributed to Kronecker (cf. [PIC] pp. 103–105; the proof below is taken from [LIM 1] pp. 229–230). We say that a simple zero p of F is *positive* if $\det(dF) > 0$ and is *negative* otherwise. Since simple zeroes are isolated, there are only finitely many of them in a compact set.

Theorem 3. *Assume that $F: U \subset \mathbf{R}^2 \to \mathbf{R}^2$ has only simple zeroes in a disk $D \subset U$ none of which is in ∂D. Then*

$$n(F; D) = P - N,$$

where P is the number of positive zeroes in D and N is the number of negative zeroes in D.

For the proof, we need the following lemma.

Lemma. *Assume that F has a unique simple zero $p \in D \subset U$. Then $n(F; D) = \pm 1$ according to $\det(dF_p) > 0$ or $\det(dF_p) < 0$.*

Proof of the Lemma. We can assume that $p = (0, 0)$. By Taylor's formula,

$$F(q) = Tq + R(q)\,|q|, \quad \lim_{q \to 0} R(q) = 0,$$

where $T = dF_p$. Consider the map $H = U \times [0,1] \to \mathbf{R}^2$

$$H(q,t) = Tq + (1-t)R(q)\,|q|\,, \quad q \in U, \quad t \in [0,1].$$

If we show that, for D sufficiently small, $H(q,t) \neq 0$, $q \in D$, we will have that $H(C(s),t)$ will be a free homotopy between the curves $F \circ C$ and $T \circ C$, hence $n(F;D) = n(T;D)$. Since T is one-to-one, the image of a circle around p will turn at most once. Thus

$$n(F;D) = n(T;D) = \pm 1,$$

as we wished to prove.

It remains to show that there exists $\varepsilon > 0$ so that $H(q,t) \neq 0$ for all $q \neq 0$, $|q| < \varepsilon$, and all $t \in [0,1]$. To do this, set $C = 1/\left|T^{-1}\right|$ and observe that, for all q,

$$|q| = \left|T^{-1}Tq\right| \leq \left|T^{-1}\right| |Tq| = 1/C\,|Tq|\,,$$

hence $|Tq| \geq C\,|q|$. Now, take $\varepsilon > 0$ so that in the disk D of radius ε we have $|R(q)| \leq C/2$ (recall that $\lim_{q\to 0} R(q) = 0$). Then, if $q \neq 0$,

$$\begin{aligned} |H(q,t)| &= |Tq + (1-t)R(q)\,|q|| \\ &\geq |Tq| - (1-t)\,|R(q)|\,|q| \geq C\,|q| - \frac{C}{2}\,|q| > 0, \end{aligned}$$

as we claimed. □

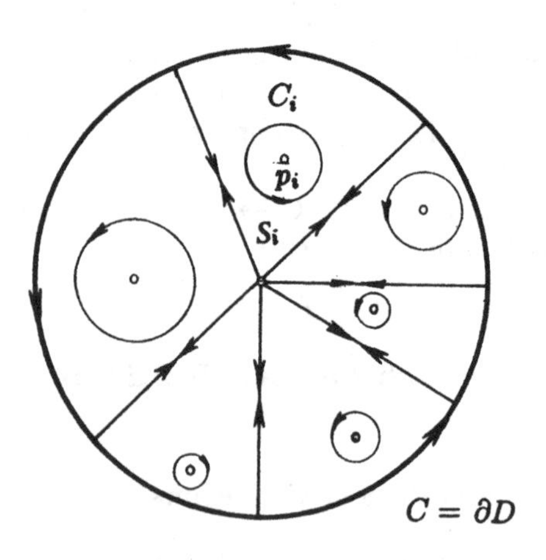

Fig. 2.2

Proof of Theorem 3. Let $p_1, \ldots, p_k$ be the zeroes of F in D. By homotopy we can change slightly D; thus we can assume that no radius of D contains more than one zero. We divide D into k sectors S_i, $i = 1, \ldots, k$, so that each such sector contains exactly one zero (Fig. 2.2). Let C_i be a circle around $p_i \in S_i$ so small that is contained in S_i. There is an obvious free homotopy taking C_i in ∂S_i. Therefore $n(F; D)$ is given by

$$\frac{1}{2\pi}\int_C \theta = \sum_i \frac{1}{2\pi}\int_{\partial S_i} \theta = \sum_i \frac{1}{2\pi}\int_{C_i} \theta = \sum_i (\pm 1) = P - N$$

as we wished to prove. □

EXERCISES

1) Show that the form $\omega = 2xy^3dx + 3x^2y^2dy$ is closed and compute $\displaystyle\int_c \omega$, where c is the arc of the parabola $y = x^2$ from (0,0) to (x, y).
2) a) Show that if ω is a differential 1–form defined in $U \subset \mathbf{R}^n$, c: $[a, b] \to U$ is a differentiable curve and $|\omega(c(t))| \leq M$, for all $t \in [a, b]$, then

$$\left|\int_c \omega\right| \leq ML,$$

where L is the length of c.

b) Let ω be a closed 1–form in $\mathbf{R}^2 - \{0\}$. Assume that ω is bounded (that is, its coefficients are bounded) in a disk of center 0. Show that ω is exact in $\mathbf{R}^2 - \{0\}$.

c) Show that the result of item (b) still holds if we only suppose that $d\omega = 0$ and that

$$\lim_{x^2+y^2\to 0} \sqrt{x^2+y^2}\ \omega = 0.$$

3) Consider the form

$$\omega = \frac{e^x}{x^2+y^2}\{(x\cos y + y\sin y)dy + (x\sin y - y\cos y)dx\}$$

defined in $\mathbf{R}^2 - \{0\}$.

a) Show that ω can be written as

$$\omega = e^x \cos y\ \omega_0 + e^x \sin y\ d(\log r),$$

where ω_0 is the element of angle at 0 and $r = \sqrt{x^2+y^2}$; check by computation that $d\omega = 0$.

b) Show that $\omega - \omega_0$ satisfies the condition of Exercise 2(c), hence is exact.

c) Compute $\int_c \omega$, where c is a simple (i.e., without self intersections) closed curve in $\mathbf{R}^2 - \{0\}$.

4) Let ω be a 1–form defined in an open set $U \subset \mathbf{R}^n$. Assume that for each closed differentiable curve c in U, $\int_c \omega$ is a rational number. Prove that ω is closed.

5) Let $U, V \subset \mathbf{R}^n$ be simply connected open sets such that $U \cap V$ is connected. Let ω be a closed 1–form so that ω is exact in U and ω is exact in V. Show that ω is exact in $U \cup V$.

6) (Applications to *complex functions*). Line integrals are quite useful in the study of complex functions $f: \mathbf{C} \to \mathbf{C}$. Here the complex plane $\mathbf{C}$ is identified with $\mathbf{R}^2$ by setting $z = x + iy$, $z \in \mathbf{C}$, $(x, y) \in \mathbf{R}^2$. It is convenient to introduce the complex differential form $dz = dx + idy$ and to write

$$f(z) = u(x, y) + iv(x, y) = u + iv.$$

Then the complex form

$$f(z)dz = (u + iv)(dx + idy) = (udx - vdy) + i(udy + vdx)$$

has $udx - vdy$ as its real part and $udy + vdx$ as its imaginary part. Define

$$\int_c f(z)dz = \int_c (udx - vdy) + i \int_c (udy + vdx)$$

Assume that u and v are C^1. Recall that f is holomorphic if and only if

$$u_x = v_y, \quad u_y = -v_x \qquad \text{(Cauchy-Riemann equations.)}$$

Show that:

a) f is holomorphic if and only if the real and imaginary parts of $f(z)dz$ are closed.

b) (Cauchy's theorem). If f is holomorphic in a simply-connected domain $U \subset \mathbf{C}$ and c is a closed curve in U, then $\int_c f(z)dz = 0$.

c) If f is holomorphic, the function $f'(z)$ (the derivative of f in z) given by the equation $df \overset{\text{def}}{=} du + idv = f'(z)dz$ is well defined and $f'(z) = u_x - iu_y$.

d) If f is holomorphic in $U \subset \mathbf{C}$ and $f'(z) \neq 0$, $z \in U$, then all zeroes of f are simple and positive; furthermore, if $D \subset U$ is a disk such that there are no zeroes in ∂D, then

$$\#\text{ of zeroes of } f \text{ in } D = \frac{1}{2\pi i} \int_{\partial D} \frac{df}{f}.$$

Hint. That the zeroes of f are simple and positive follows from

$$\det(df) = u_x^2 + u_y^2 = |f'(z)|^2 > 0.$$

Now, a computation shows that

$$\frac{df}{f} = \frac{du + idv}{u + iv} = \frac{udv + vdu}{u^2 + v^2} + i\frac{udv - vdu}{u^2 + v^2}$$
$$= \frac{1}{2}d(\log(u^2 + v^2)) + i\frac{udv - vdu}{u^2 + v^2}.$$

Thus, by Kronecker's formula,

$$\frac{1}{2\pi i}\int_{\partial D}\frac{df}{f} = \frac{1}{2\pi}\int\frac{udv - vdu}{u^2v^2} = \# \text{ of zeroes of } f \text{ in } D.$$

7) Consider the form

$$\omega = \frac{2(x^2 - y^2 - 1)dy - 4xydx}{(x^2 + y^2 - 1)^2 + 4y^2}$$

defined in $\mathbf{R}^2 - \{p_1 \cup p_2\}$, $p_1 = (1,0)$, $p_2 = (-1,0)$. Let D_1 and D_2 be disks centered in p_1 and p_2, respectively, and so small that $p_2 \notin D_1$, $p_1 \notin D_2$.

a) Show that the integrals

$$\frac{1}{2\pi}\int_{\partial D_1}\omega = +1 \qquad \frac{1}{2\pi}\int_{\partial D_2}\omega = -1,$$

where ∂D_1 and ∂D_2 are oriented counterclockwise.
Hint. Set $F = (f,g) = (x^2 + y^2 - 1, 2y)$ and observe that $\omega = \frac{fdg - gdf}{f^2 + g^2}$. Notice that p_1 and p_2 are the only zeroes of F, where p_1 is a positive zero and p_2 is a negative zero.

b) Conclude by homotopy that the integral of ω along the curve C below (Fig. 2.3), with the indicated orientation, is 4π.

8. Show that if $F: U \subset \mathbf{R}^2 \to \mathbf{R}^2$ satisfies $F(-q) = -F(q)$ for all $q \in D \subset U$, where D is a disk centered at (0,0), and F has no zeroes in ∂D, then $n(F; D)$ is an odd integer; in particular $n(F; D) \neq 0$, and there exists a zero of F in D.

9. (*Line integrals of vector fields*). Let v be a differentiable vector field defined in an open set $U \subset \mathbf{R}^n$. At the end of Chapter 1, we have associated to v a differentiable 1-form ω by $\omega(u) = \langle v, u\rangle$, for all $u \in \mathbf{R}^n$. Let $c: [a,b] \to U$ be a piecewise differentiable curve. By defining

$$\int_c v = \int_c \omega,$$

we can translate the results of the present chapter into properties of integrals of vector fields along curves. For instance:

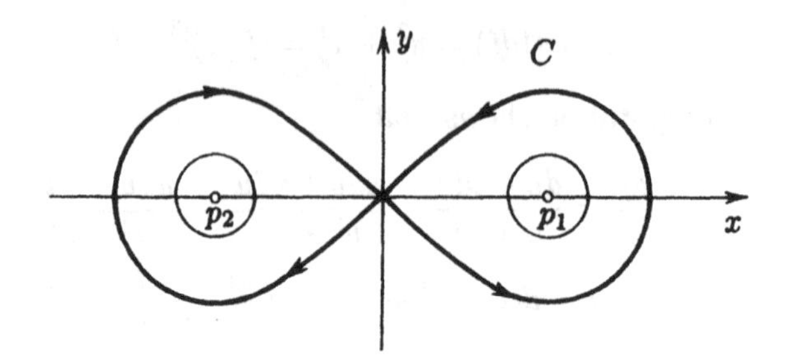

Fig. 2.3

a) Assume that $n = 3$, that $U \subset \mathbf{R}^3$ is simply-connected, and that rot $v = 0$ (see the definition of rot v in Exercise 14, Chapter 1). Prove that there exists a function $f: U \to \mathbf{R}$ such that $v = \text{grad } f$ (see Exercise 12, Chapter 1); furthermore for any curve c joining the points $p_1, p_2 \in U$,

$$\int_c v = f(p_2) - f(p_1)$$

(If v is a field of forces, f or rather $-f$, is called the potential energy of v and the expression above means that the work done by v along c is equal to change of potential energy between p_1 and p_2).

b) Let the situation be as in (a) and assume, in addition, that div $v = 0$ (see Exercise 11, Chapter 1). Show that the potential is then a harmonic function, i.e., $f_{xx} + f_{yy} + f_{zz} = 0$.

10. Let ω be a differentiable 1-form defined in $U \subset \mathbf{R}^2$. A *local integrating factor* at p for ω is a function $g: V \to \mathbf{R}$ defined in a neighborhood $V \subset U$ of p such that the form $g\omega$ is exact in V, i.e., there exists a function $f: V \to \mathbf{R}$ with $g\omega = df$.

a) Show that if $\omega(p) \neq 0$ there exists a local integrating factor at p.
Hint. The condition $\omega(v) = 0$ determines a vector field v, in a neighborhood of p, which is nowhere zero. By the fundamental theorem of ordinary differential equations, there exists a neighborhood V of p and a function $f: V \to \mathbf{R}$ (the so-called first integral of f) so that $f = \text{const.}$ along the trajectories of v. Thus

$$df(v) = 0 = \omega(v).$$

It follows that $df = g\omega$.

b) Prove that if $g: V \to \mathbf{R}$ is a local integrating factor at $p \in V$, i.e., $df = g\omega$ and $\theta: \mathbf{R} \to \mathbf{R}$ is any differential function, then $\tilde{g}: V \to \mathbf{R}$ defined by $\tilde{g}(p) = d\theta(f(p)) \cdot g(p)$ is still an integrating factor.

Hint. We use the notation of the hint in (a). $\theta(f)$ is still constant along the trajectories of v. Thus $d(\theta(f)) = \tilde{g}\omega$ or $d\theta \cdot g\omega = \tilde{g}\omega$, where $\tilde{g}$ is a new integrating factor. Since $\omega \neq 0$, $d\theta \cdot g = \tilde{g}$.

3. Differentiable Manifolds

Differential forms were introduced in the first chapter as objects in $\mathbf{R}^n$; however, they, as everything else that refers to differentiability, live naturally in a differentiable manifold, a concept that we will develop presently.

We will start with the most familiar example of a differentiable manifold, namely a regular surface in $\mathbf{R}^3$. In what follows, we will use as a reference "M. do Carmo, Differential Geometry of Curves and Surfaces, Prentice-Hall, 1976", to be quoted as [**dC**]. Let us recall (cf. [**dC**], Chap. 2 §2.2) that a subset $S \subset \mathbf{R}^3$ is a regular surface if, for each $p \in S$, there exist a neighborhood V of p in $\mathbf{R}^3$ and a map $f_\alpha: U_\alpha \subset \mathbf{R}^2 \to V \cap S$ of an open set U_α in $\mathbf{R}^2$ onto $V \cap S$ such that:

1) f_α is a differentiable homeomorphism.
2) the differential $(df_\alpha)_q: T_q(U_\alpha) \to \mathbf{R}^3$ is injective for each $q \in U_\alpha$.

The map $f_\alpha: U_\alpha \to S$ is called a parametrization of S around p.

The most important consequence of the above definition is the fact that the change of parameters is a diffeomorphism. More precisely, if $f_\alpha: U_\alpha \to S$ and $f_\beta: U_\beta \to S$ are two parametrizations such that $f_\alpha(U_\alpha) \cap f_\beta(U_\beta) = W \neq \phi$, then (cf. [**dC**], Theorem of §2.5) the maps

$$f_\beta^{-1} \circ f_\alpha: f_\alpha^{-1}(W) \to \mathbf{R}^2,$$
$$f_\alpha^{-1} \circ f_\beta: f_\beta^{-1}(W) \to \mathbf{R}^2$$

are differentiable. It follows that on a regular surface it makes sense to talk about differentiable functions and to apply the methods of Differential Calculus.

The most serious problem with the above definition is its dependence on $\mathbf{R}^3$. Indeed the natural idea of an abstract surface is that of an object that is, in a certain sense, *two*-dimensional, and to which we can apply, locally, the Differential Calculus in $\mathbf{R}^2$. Such an idea of an abstract surface (i.e., with no reference to an ambient space) has been foreseen since Gauss. It took, however, about a century for the definition to reach the definitive form that we present below. One of the reasons for this delay was that, even for surfaces in $\mathbf{R}^3$, the fundamental rôle of the change of parameters was not clearly understood (see Remark 1).

Since nothing is gained in limiting ourselves to dimension two, we will present the definition for dimension n.

Definition 1. An *n-dimensional differentiable manifold* is a set M together with a family of injective maps $f_\alpha: U_\alpha \subset \mathbf{R}^n \to M$ of open sets U_α in $\mathbf{R}^n$ into M such that:

1) $\bigcup_\alpha f_\alpha(U_\alpha) = M$.

2) For each pair α, β, with $f_\alpha(U_\alpha) \cap f_\beta(U_\beta) = W \neq \phi$, the sets $f_\alpha^{-1}(W)$ and $f_\beta^{-1}(W)$ are open sets in $\mathbf{R}^n$ and the maps $f_\beta^{-1} \circ f_\alpha$, $f_\alpha^{-1} \circ f_\beta$ are differentiable (Fig. 3.1).

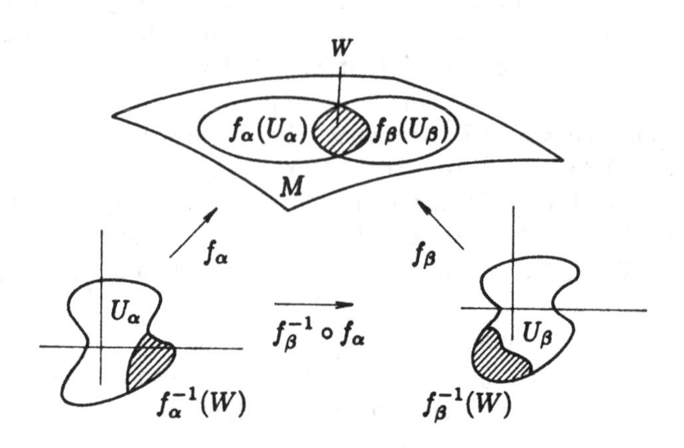

Fig. 3.1

3) The family $\{(U_\alpha, f_\alpha)\}$ is maximal relative to (1) and (2).

The pair (U_α, f_α) with $p \in f_\alpha(U_\alpha)$ is called a *parametrization* (or a *coordinate system*) of M at p; $f_\alpha(U_\alpha)$ is then called a *coordinate neighborhood* of p. A family (f_α, U_α) satisfying the properties (1) and (2) is called a *differentiable structure* on M.

Condition 3 is a technical condition. Actually, we can always extend a differentiable structure into a maximal one, by adjoining to the given structure all parametrizations that together with some parametrization of the given structure satisfy condition 2. Thus, with a certain abuse of language, we can say that a differentiable manifold is a set endowed with a differentiable structure, the extension to the maximal one being assumed whenever needed.

Remark 1. A comparison between the definition of a regular surface in $\mathbf{R}^3$ and the definition of a differentiable manifold shows that the crucial point was to introduce the fundamental property of change of parameters (which is a theorem for surfaces) as an axiom in Definition 1. As we will soon see, this is what allows to transport to differentiable manifolds all the notions of Differential Calculus in $\mathbf{R}^n$.

Remark 2. A differentiable structure on a set M induces in a natural way a topology in M. It suffices to define that $A \subset M$ is an *open set* if $f_\alpha^{-1}(A \cap f_\alpha(U_\alpha))$ is an open set in $\mathbf{R}^n$, for all α. It is easily checked that this defines a topology in M in which the sets $f_\alpha(U_\alpha)$ are open and the maps f_α are continuous.

Remark 3. The natural topology of a differentiable manifold can be quite strange. In particular, it can happen that one (or both) of the following axioms do not hold:

a) *Axiom of Hausdorff.* Given two distinct points of M there exist neighborhoods of these points that do not intersect.
b) *Axiom of countable basis.* M can be covered by a countable number of coordinate neighborhoods (we say then that *M has a countable basis*)

Axiom (a) is essential to prove that the limit of a converging sequence is unique, and axiom (b) is essential for the existence of a partition of unity (cf. Chapter 3) which is an almost indispensable tool for the study of the topology of manifolds.

We shall assume, from now on, that all the manifolds to be considered are Hausdorff and have a countable basis.

Of course, a regular surface is an example of a 2-dimensional differentiable manifold. A trivial example of an n-dimensional differentiable manifold is the euclidean space $\mathbf{R}^n$ with the differentiable structure given by the identity. Less trivial examples are the following ones.

Example 1. The real projective plane $P^2(\mathbf{R})$. We will denote by $P^2(\mathbf{R})$ the set of all lines in $\mathbf{R}^3$ that pass through the origin $(0,0,0)$ of $\mathbf{R}^3$, i.e., $P^2(\mathbf{R})$ is the set of "directions" in $\mathbf{R}^3$.

We want to introduce a differentiable structure in $P^2(\mathbf{R})$. For that, let $(x,y,z) \in \mathbf{R}^3$, and notice that $P^2(\mathbf{R})$ is the quotient space of $\mathbf{R}^3 - \{(0,0,0)\}$ by the equivalence relation $\sim$:

$$(x,y,z) \sim (\lambda x, \lambda y, \lambda z), \qquad \lambda \in \mathbf{R}, \quad \lambda \neq 0.$$

The points of $P^2(\mathbf{R})$ will be denoted by $[x,y,z]$.

We now define sets V_1, V_2, V_3 in $P^2(\mathbf{R})$ by:

$$V_1 = \{[x,y,z]; x \neq 0\},$$

$$V_2 = \{[x,y,z]; y \neq 0\},$$

$$V_3 = \{[x,y,z]; z \neq 0\},$$

and maps $f_i: \mathbf{R}^2 \to V_i$, $i = 1, 2, 3$, by

$$f_1(u, v) = [1, u, v], \quad f_2(u, v) = [u, 1, v], \quad f_3(u, v) = [u, v, 1],$$

where $(u, v) \in \mathbf{R}^2$. Geometrically V_2, for instance, is the set of lines of $\mathbf{R}^3$ that pass through the origin and do not belong to the plane xOz. We claim that the family $\{(f_i, \mathbf{R}^2)\}$ is a differentiable structure for $P^2(\mathbf{R})$.

Each f_i, $i = 1, 2, 3$, is clearly bijective and $\bigcup_i f_i(\mathbf{R}^2) = P^2(\mathbf{R})$. It remains to be shown that $f_i^{-1}(V_i \cap V_j)$ is open in $\mathbf{R}^2$ and $f_j^{-1} \circ f_i$ is differentiable. Let us consider the case $i = 1$, $j = 2$; the other cases are entirely analogous.

The points of $f_1^{-1}(V_1 \cap V_2)$ are of the form (u, v) with $u \neq 0$. Thus $f_1^{-1}(V_1 \cap V_2)$ is open in $\mathbf{R}^2$, and

$$f_2^{-1} \circ f_1(u, v) = f_2^{-1}([1, u, v]) = f_2^{-1}([\frac{1}{u}, 1, \frac{v}{u}]) = (\frac{1}{u}, \frac{v}{u})$$

is clearly differentiable, as we claimed.

Example 1'. (The real projective space). Example 1 can easily be generalized. Let $(x_1, \ldots, x_{n+1}) \in \mathbf{R}^{n+1}$ and define the n-dimensional *real projective space* $P^n(\mathbf{R})$ as the quotient space of $\mathbf{R}^{n+1} - \{0\}$ by the equivalence relation:

$$(x_1, \ldots, x_{n+1}) \sim (\lambda x_1, \ldots, \lambda x_{n+1}), \quad \lambda \in \mathbf{R}, \quad \lambda \neq 0;$$

points in $P^n(\mathbf{R})$ will be denoted by $[x_1, \ldots, x_{n+1}]$.

Define subsets $V_i \subset P^n(\mathbf{R})$, $i = 1, \ldots, n+1$, by

$$V_i = \{[x_1, \ldots, x_{n+1}]; x_i \neq 0\},$$

and maps $f_i: \mathbf{R}^n \to V_i$ by

$$f_i(y_1, \ldots, y_n) = [y_1, \ldots, y_{i-1}, 1, y_i, \ldots, y_n].$$

Proceeding as in Example 1, one easily checks that the family $\{(f_i, \mathbf{R}^n)\}$ is a differentiable structure in $P^n(\mathbf{R})$. (See Exercise 1).

Example 2. (The Klein bottle). The Klein bottle is the subset of $\mathbf{R}^4$ defined as follows (Fig. 3.2). Let Ox, Oy, Oz, Ow be the four coordinate axis in $\mathbf{R}^4$. Let S be a circle with radius r, contained in the plane xOz, with center C in the axis Ox so that C is at a distance $a > r$ from 0. The Klein bottle is generated by rotating this circle around Oz in such a way that when the center C has described a rotation of an angle u in the plane xOy, the plane of S has described a rotation of angle $u/2$ around OC in the 3-space $OCOzOw$ (this is possible because we are in $\mathbf{R}^4$).

Let u and v be as in Fig. 3.1, and let $U_1 \subset \mathbf{R}^2$ be given by

$$U_1 = \{(u, v) \in \mathbf{R}^2; 0 < u < 2\pi, 0 < v < 2\pi\}$$

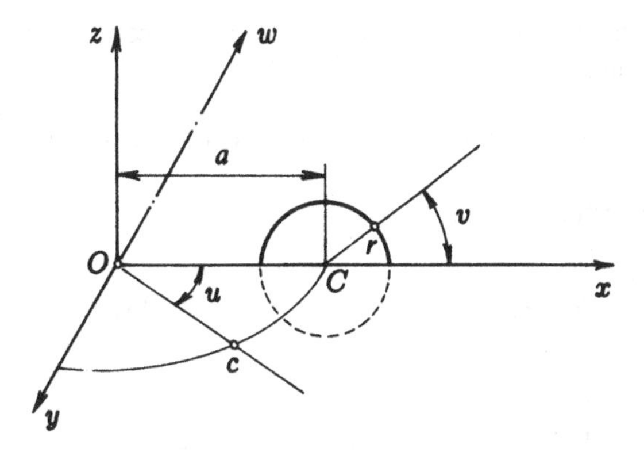

Fig. 3.2

Now define a map $f_1: U_1 \to \mathbf{R}^4$ by

$$f_1(u,v) = \begin{cases} x = & (r\cos v + a)\cos u \\ y = & (r\cos v + a)\sin u \\ z = & r\sin v\cos u/2 \\ w = & r\sin v\sin u/2. \end{cases}$$

It is clear that $f_1(U_1)$ contains the points of the Klein bottle that are not on the circles $u = 0$ and $v = 0$. We claim that f_1 is injective.

To see that, let us first assume that $z \neq 0$. Then both $\sin v$ and $\cos u/2$ are nonzero. Since $0 < u/2 < \pi$, the expression $\frac{w}{z} = \tan\frac{u}{2}$ determines u. By knowing u, we can use that

$$\sin v = \frac{w}{r\sin\frac{u}{2}}, \qquad \cos v = \frac{+\sqrt{x^2+y^2} - a}{r}$$

to determine v. This proves our claim if $z \neq 0$. If $z = 0$ then $v = \pi$ or $u = \pi$, and again injectivity is easily checked.

It can be shown that if we change the origin of u and the origin of v, we can cover the whole Klein bottle by the images of maps similar to the above. For instance, if we define a map $f_2: U_2 \to \mathbf{R}^4$ given by

$$f_2 \begin{cases} x = & -(r\cos\overline{v} + a)\sin\overline{u} \\ y = & (r\cos\overline{v} + a)\cos\overline{u} \\ z = & r\sin\overline{v}\cos\left(\frac{\overline{u}}{2} + \frac{\pi}{4}\right) \\ w = & r\sin\overline{v}\sin\left(\frac{\overline{u}}{2} + \frac{\pi}{4}\right) \end{cases}, \qquad (\overline{u},\overline{v}) \in U_2,$$

(geometrically, this means that we measure $\overline{u}$ from Oy), we see that $f_2(U_2)$ includes the points of the Klein bottle with $u = 0$. It is easily checked that f_2

is injective. Notice that $f_1(U_1) \cap f_2(U_2) = W$ is not connected but has two connected components:

$$W_1 = \{f_1(u,v); \frac{\pi}{2} < u < 2\pi\}, \quad W_2 = \{f_1(u,v); 0 < u < \frac{\pi}{2}\}.$$

The change of coordinates is given by:

$$f_2^{-1} \circ f_1 \begin{cases} \overline{u} = & u - \pi/2 \\ \overline{v} = & v \end{cases} \qquad \text{in } W_1,$$

$$f_2^{-1} \circ f_1 \begin{cases} \overline{u} = & u + 3\pi/2 \\ \overline{v} = & 2\pi - v \end{cases} \qquad \text{in } W_2,$$

which is clearly differentiable.

In a similar way, we can find an injective map $f_3: U_3 \to \mathbf{R}^4$ whose image covers the Klein bottle minus the circle $v = 0$ (this amounts to change the origin of v). It also is not difficult to check that the changes of coordinates $f_j^{-1} \circ f_i$, $i, j = 1, 2, 3$, are differentiable, and this shows that the family (f_i, U_i) is a differentiable structure on the Klein bottle.

Remark 4. The Klein bottle can be thought of as a twisted torus in the following sense. The torus is obtained from a rectangle by identifying the opposite sides. In the Klein bottle, one of the sides is reflected across its center before performing the identification (Fig. 3.3). It can be shown that the Klein bottle is not a regular surface in $\mathbf{R}^3$: the model of Fig. 3.3 has self-intersections.

Before presenting further examples, we need to extend to differentiable manifolds the local Differential Calculus that holds for $\mathbf{R}^n$.

From now on, when we denote a differentiable manifold by M^n, the upper index n will denote the dimension of $M = M^n$.

Definition 2. Let M_1^n and M_2^m be differentiable manifolds. A map $\varphi: M_1 \to M_2$ is *differentiable at a point* $p \in M_1$ if given a parametrization $g: V \subset \mathbf{R}^m \to M_2$ around $\varphi(p)$, there exists a parametrization $f: U \subset \mathbf{R}^n \to M_1$ around p such that $\varphi(f(U)) \subset g(V)$ and the map

$$g^{-1} \circ \varphi \circ f: U \subset \mathbf{R}^n \to \mathbf{R}^m$$

is differentiable at $f^{-1}(p)$. The map φ is *differentiable in an open set* of M_1 if it is differentiable at all points of this set.

The map $g^{-1} \circ \varphi \circ f$ is the *expression* of φ in the parametrizations f and g. Since the change of parameters is differentiable, the fact that φ is differentiable does not depend on the choice of parametrizations.

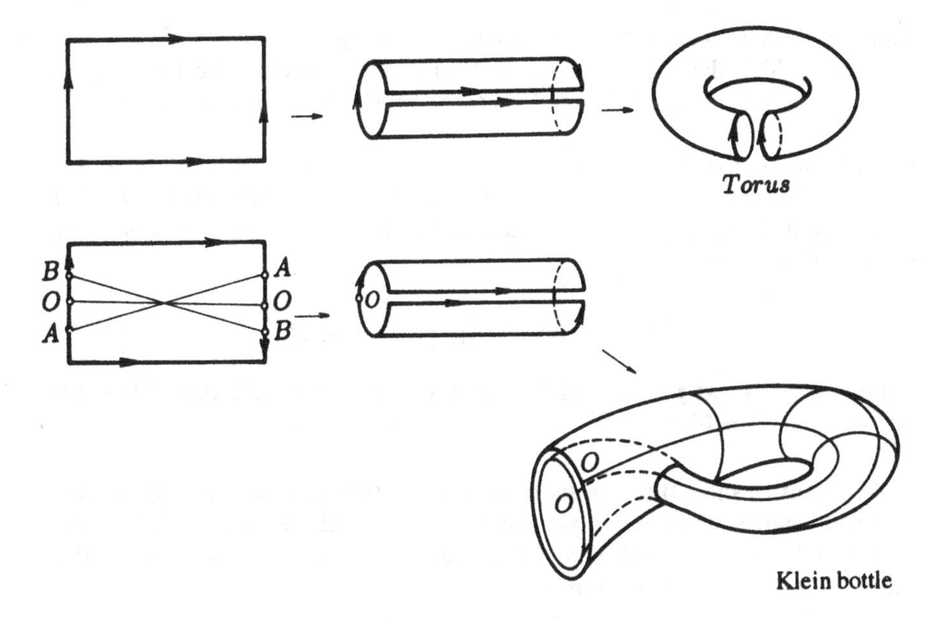

Fig. 3.3

In particular, it follows from the above that we can talk about differentiable functions ($\varphi: M^n \to \mathbf{R}$) and differentiable curves ($\varphi: I \subset \mathbf{R} \to M^n$) on a differentiable manifold ($I \subset \mathbf{R}$ will always denote an open interval of the real line containing the origin $0 \in \mathbf{R}$).

Now we would like to define the notion of a tangent vector to a differentiable curve on a differentiable manifold. For the case of a differentiable curve $\alpha: I \subset \mathbf{R} \to S \subset \mathbf{R}^3$ on a regular surface in $\mathbf{R}^3$, the tangent vector $\alpha'(t)$ is merely the speed of α as a vector in $\mathbf{R}^3$. Since we do not have the support of a nice ambient space, we must choose a characteristic property of the tangent vector that does not depend on the ambient space.

For that, let $\alpha: (-\varepsilon, \varepsilon) \to \mathbf{R}^n$ be a differentiable curve in $\mathbf{R}^n$, with $\alpha(0) = p \in \mathbf{R}^n$, and write

$$\alpha(t) = (x_1(t), \dots, x_n(t)), \quad t \in (-\varepsilon, \varepsilon), \quad (x_1, \dots, x_n) \in \mathbf{R}^n.$$

Then $\alpha'(0) = (x_1'(0), \dots, x_n'(0)) = v \in \mathbf{R}^n$. Now let φ be a real function in $\mathbf{R}^n$, differentiable in a neighborhood of p. Then the derivative of φ along v at p is given by

$$\frac{d}{dt}(\varphi \circ \alpha)\Big|_{t=0} = \sum_{i=1}^{n} \frac{\partial \varphi}{\partial x_i} \frac{dx_i}{dt}\Big|_{t=0} = \left(\sum_{i=1}^{n} x_i'(0) \frac{\partial}{\partial x_i} \right) \varphi.$$

Thus the "directional derivative along v" is an operator over differentiable functions which depends only on v. This is the characteristic property of vectors that we will use to extend them to differentiable manifolds.

Definition 3. Let $\alpha: I \to M$ be a differentiable curve on a differentiable manifold M, with $\alpha(0) = p \in M$, and let D be the set of functions of M which are differentiable at p. *The tangent vector* to the curve α at p is the map $\alpha'(0): D \to \mathbf{R}$ given by

$$\alpha'(0)\varphi = \frac{d}{dt}(\varphi \circ \alpha)\big|_{t=0}, \qquad \varphi \in D.$$

A tangent vector at $p \in M$ is the tangent vector of some differentiable curve $\alpha: I \to M$ with $\alpha(0) = p$.

We want to show that the set of tangent vectors at a point $p \in M^n$ makes up an n-dimensional real vector space. For that, choose a parametrization $f: U \subset \mathbf{R}^n \to M$ around $p = f(0, \ldots, 0)$. Then a curve $\alpha: I \to M$ and a function $\varphi \in D$ can be written as:

$$f^{-1} \circ \alpha(t) = (x_1(t), \ldots, x_n(t)),$$
$$\varphi \circ f(q) = \varphi(x_1, \ldots, x_n), \qquad q = (x_1, \ldots, x_n) \in U,$$

respectively. Thus

$$\begin{aligned}\alpha'(0)\varphi &= \frac{d}{dt}(\varphi \circ \alpha)\big|_{t=0} = \frac{d}{dt}\varphi(x_1(t), \ldots, x_n(t))\big|_{t=0} \\ &= \left(\sum_{i=1}^{n} x_i'(0) \left(\frac{\partial}{\partial x_i}\right)_0\right)\varphi,\end{aligned}$$

so that the tangent vector $\alpha'(0)$ at p can be written as

$$\alpha'(0) = \sum_{i=1}^{n} x_i'(0) \left(\frac{\partial}{\partial x_i}\right)_0. \tag{1}$$

Notice that $\left(\frac{\partial}{\partial x_i}\right)_0$ is the tangent vector at p to the "coordinate curve"

$$x_i \to f(0, \ldots, 0, x_i, 0, \ldots, 0).$$

Now let T_f be the vector space generated by $\left\{\left(\frac{\partial}{\partial x_i}\right)_0\right\}$, $i = 1, \ldots, n$.

Lemma 1. *The set T_pM of tangent vectors to M at p is equal to T_f.*

Proof. We have just shown that $T_pM \subset T_f$. Conversely, if $v \in T_f$, then $v = \sum_i \lambda_i \left(\frac{\partial}{\partial x_i}\right)_0$. Let $\alpha: I \to M$ be given in the parametrization f by $x_i = \lambda_i t$. Then $\alpha'(0) = v$, i.e., $v \in T_pM$. □

It follows that T_pM is a vector space. Furthermore, the choice of a parametrization f determines a basis $\left\{\left(\frac{\partial}{\partial x_i}\right)_0\right\}$ for T_pM. Thus T_pM is an n-dimensional vector space which is called the *tangent space* of M at p. The basis $\left\{\left(\frac{\partial}{\partial x_i}\right)_0\right\}$ is called the *associated basis* to the parametrization f.

With the notion of a tangent space we can define the notions of differential of a map $\varphi: M_1^n \to M_2^m$

Definition 4. Let M_1^n and M_2^m be differentiable manifolds and let $\varphi: M_1 \to M_2$ be a differentiable map. For each $p \in M$, *the differential of* φ *at* p is the linear map $d\varphi_p: T_pM_1 \to T_{\varphi(p)}M_2$ which associates to each $v \in T_pM_1$ the vector $d\varphi_p(v) \in T_{\varphi(p)}M_2$ defined as follows: Choose a differentiable curve $\alpha: (-\varepsilon, \varepsilon) \to M_1$, with $\alpha(0) = p$, $\alpha'(0) = v$; then $d\varphi_p(v) = (\varphi \circ \alpha)'(0)$.

For the definition to make sense, it is necessary to show that $(\varphi \circ \alpha)'(0)$ does not depend on the choice of α and that $d\varphi_p$ is in fact a linear map. This is proved by taking parametrizations around p and $\varphi(p)$ thus reducing the question to the case of a map from $\mathbf{R}^n$ to $\mathbf{R}^m$, where the properties are well known (or can easily be proved); for the case $n = 2$, $m = 3$, see, for instance, [dC] §2.4.

Definition 5. Let M_1 and M_2 be differentiable manifolds. A map $\varphi: M_1 \to M_2$ is a *diffeomorphism* if it is differentiable, bijective, and its inverse φ^{-1} is also differentiable. The map φ is a *local diffeomorphism* at $p \in M$ if there exist neighborhoods U of p and V of $\varphi(p)$ such that $\varphi: U \to V$ is a diffeomorphism.

The linear map $d\varphi_p$ may be thought as a first order approximation of the map φ around p. Probably the most important local theorem in Calculus is *the inverse function theorem* which states that *if* $d\varphi_p$ *is an isomorphism then* φ *is a local diffeomorphism at* p. Being a local theorem, it extends immediately to differentiable manifolds.

Example 3. (The tangent bundle). Let M^n be a differentiable manifold and let

$$TM = \{(p, v); p \in M, v \in T_pM\},$$

i.e., TM is the set of all tangent vectors to M. We will introduce in TM a differentiable structure (of dimension 2n); with such an structure, TM is called the *tangent bundle* of M.

Let $f_\alpha: U_\alpha \subset \mathbf{R}^n \to M$ be a parametrization of M with $(x_1^\alpha, \ldots, x_n^\alpha) \in U_\alpha$. For $w \in T_{f_\alpha(q)}M$, $q \in U_\alpha$, we can write

$$w = \sum_i y_i^\alpha \frac{\partial}{\partial x_i^\alpha}.$$

Define a map $F_\alpha: U_\alpha \times \mathbf{R}^n \to TM$ by

$$F_\alpha(x_1^\alpha,\dots,x_n^\alpha,y_1^\alpha,\dots,y_n^\alpha)=\left(f_\alpha(x_1^\alpha,\dots,x_n^\alpha),\sum_i y_i^\alpha\frac{\partial}{\partial x_i^\alpha}\right)$$

We claim that if $\{(U_\alpha,f_\alpha)\}$ is a differentiable structure for M then $\{(U_\alpha\times\mathbf{R}^n,F_\alpha)\}$ is a differentiable structure for TM. Geometrically, this means that we take as coordinates of a point $(p,v)\in TM$ the coordinates of p together with the coordinates of v in the associated basis.

Let

$$(p,v)\in F_\alpha(U_\alpha\times\mathbf{R}^n)\cap F_\beta(U_\beta\times\mathbf{R}^n),$$

that is,

$$(p,v)=(f_\alpha(q_\alpha),df_\alpha(v_\alpha))=(f_\beta(q_\beta),df_\beta(v_\beta)),$$

where $q_\alpha\in U_\alpha$, $q_\beta\in U_\beta$, $v_\alpha,v_\beta\in\mathbf{R}^n$. Then

$$\begin{aligned}F_\beta^{-1}\circ F_\alpha(q_\alpha,v_\alpha)&=F_\beta^{-1}(f_\alpha(q_\alpha),df_\alpha(v_\alpha))\\&=(f_\beta^{-1}\circ f_\alpha(q_\alpha),d(f_\beta^{-1}\circ f_\alpha)(v_\alpha)).\end{aligned}$$

Since $f_\beta^{-1}\circ f_\alpha$ is differentiable, so is $d(f_\beta^{-1}\circ f_\alpha)$. It follows that $F_\beta^{-1}\circ F_\alpha$ is differentiable. This proves condition (2) of Definition 1. Since condition (1) is immediate, our claim follows.

Definition 6. Let M^m and N^n be differentiable manifolds. A differentiable map $\varphi: M\to N$ is an *immersion* if $d\varphi_p: T_pM\to T_{\varphi(p)}N$ is injective for all $p\in M$. If, in addition, φ is a homeomorphism onto $\varphi(M)\subset N$, where $\varphi(M)$ has the topology induced by N, φ is an *embedding*. If $M\subset N$ and the inclusion $i: M\subset N$ is an embedding, we say that M is a *submanifold* of N.

Example 4.

a) The curve $\alpha:\mathbf{R}\to\mathbf{R}^2$ given by $\alpha(t)=(t^3,t^2)$ is a differentiable map but not an immersion. In fact, the condition of immersion is this case is equivalent to the fact that $\alpha'(t)\neq 0$, which does not hold for $t=0$.

b) The curve $\alpha(t)=(t^3-4t,t^2-4)$ is an immersion that is not an embedding, since it has a self-intersection at $(0,0)$ (for $t=2$, $t=-2$).

c) The curve (Fig. 3.4)

$$\alpha(t)\begin{cases}=(0,-(t+2)), & t\in(-3,-1)\\ =\text{a regular curve as in Fig. 3.2}, & t\in(-1,-1/\pi)\\ =(-t,-\sin(1/t)), & t\in(-1/\pi,0)\end{cases}$$

is an immersion $\alpha:(-3,0)\to\mathbf{R}^2$ without self-intersections. However, α is not an embedding. For, in the topology of $\mathbf{R}^2$, a neighborhood of a point p in the vertical part of the curve has infinitely many connected components, where as in the topology induced by α it is a connected interval.

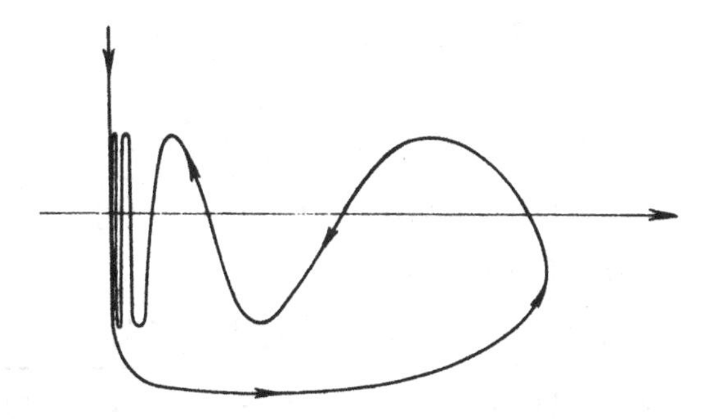

Fig. 3.4

d) The inclusion $i: S^2 \subset \mathbf{R}^3$ of a regular surface in $\mathbf{R}^3$ is an embedding. This follows from conditions (1) and (2) of the definition of a regular surface in the beginning of this chapter.

Example 5. (Regular surfaces of $\mathbf{R}^n$*).* The natural generalization of the idea of a regular surface in $\mathbf{R}^3$ is the notion of a k-dimensional surface in $\mathbf{R}^n$, $k < n$. A subset $M^k \subset \mathbf{R}^n$ is a k-dimensional *regular surface* if for each $p \in M$ there exist a neighborhood V of p in $\mathbf{R}^n$ and a map $f: U \subset \mathbf{R}^k \to M \cap V$ of an open set U of $\mathbf{R}^k$ onto $M \cap V$ such that:

1) f is a differentiable homeomorphism,
2) $(df)_q: \mathbf{R}^k \to \mathbf{R}^n$ is injective for all $q \in U$.

Except for the dimensions involved, the definition is the same as the one given for a regular surface of $\mathbf{R}^3$.

It can be proved that given two parametrizations $f_1: U_1 \subset \mathbf{R}^k \to M$, $f_2: U_2 \subset \mathbf{R}^k \to M$, with $f_1(U_1) \cap f_2(U_2) = W \neq \phi$, the change of parameters $f_1^{-1} \circ f_2: f_2^{-1}(W) \to f_1^{-1}(W)$ is a diffeomorphism. The proof is the same as for surfaces (see [**dC**] pg. 71).

Example 6. It can be shown that the maps f_i, $i = 1, 2, 3$, given in Example 2, have injective differentials. This shows that the Klein bottle is a 2-surface of $\mathbf{R}^4$.

Example 7. (The projective plane revisited). The set of straight lines of $\mathbf{R}^3$ that pass through the origin may be looked upon as the quotient space of the unit sphere $S^2 = \{p \in \mathbf{R}^3; |p| = 1\}$ by the equivalence relation that identifies p with the antipodal point $-p$. For each straight line through the origin determines in S^2 two antipodal points, and the correspondence thus obtained is clearly bijective.

Taking this into account, let us introduce another differentiable structure in $P^2(\mathbf{R})$. For that, notice that we can cover the regular surface $S^2 \subset \mathbf{R}^3$ by a family of parametrizations:

$$f_i^+ : U_i \to S^2, \quad f_i^- : U_i \to S^2, \quad i = 1, 2, 3,$$

where

$$U_1 = \{(x_1, x_2, x_3) \in \mathbf{R}^3; x_1 = 0, x_2^2 + x_3^2 < 1\}$$
$$f_1^+(x_2, x_3) = (D_1, x_2, x_3), \qquad D_1 = \sqrt{1 - (x_2^2 + x_3^2)}$$
$$f_1^-(x_2, x_3) = (-D_1, x_2, x_3);$$
$$U_2 = \{(x_1, x_2, x_3) \in \mathbf{R}^3; x_2 = 0, x_1^2 + x_3^2 < 1\}$$
$$f_2^+(x_1, x_3) = (x_1, D_2, x_3), \qquad D_2 = \sqrt{1 - (x_1^2 + x_3^2)}$$
$$f_2^-(x_1, x_3) = (x_1, -D_2, x_3);$$

and similarly for $i = 3$ (Fig. 3.5). It is immediate to check that this is in fact a differentiable structure for S^2.

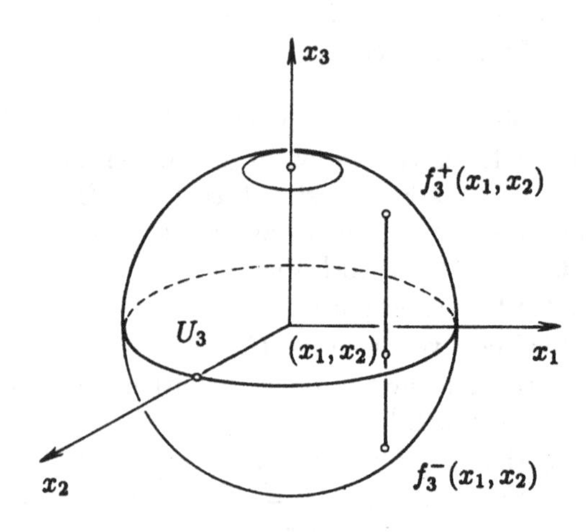

Fig. 3.5

Let $\pi: S^2 \to P^2(\mathbf{R})$ be the canonical projection, that is, $\pi(p) = \{p, -p\}$ and notice that $\pi(f_i^+(U_i)) = \pi(f_i^-(U_i))$. We will define $g_i: U_i \to P^2(\mathbf{R})$ by

$$g_i = \pi \circ f_i^+$$

Since the restriction of π to $f_i^+(U_i)$ is injective, we obtain that

$$g_i^{-1} \circ g_j = (\pi \circ f_i^+)^{-1} \circ (\pi \circ f_j^+) = (f_i^+)^{-1} \circ f_j^+.$$

It follows that $g_i^{-1} \circ g_j$ is differentiable for all $i, j = 1, 2, 3$, hence $\{(U_i, g_i)\}$ is a differentiable structure for $P^2(\mathbf{R})$.

Actually the above structure and that of Example 1 will determine the same maximal structure. For the coordinate neighborhoods are the same and the coordinate changes are given (for $i = 1$, say) by

$$\left(1, \frac{x_2}{x_1}, \frac{x_3}{x_1}\right) \longleftrightarrow (D, x_2, x_3)$$

which is clearly differentiable. Notice also that it follows from the above that the canonical projection $\pi: S^2 \to P^2(\mathbf{R})$ is a local diffeomorphism at each point of S^2.

Remark 5. It is possible to prove that the projective plane cannot be embedded in $\mathbf{R}^3$. Fig. 3.6 describes a differentiable map of the projective plane into $\mathbf{R}^3$ that has a line segment of self-intersection and two singular points at the endpoints of this segment. With a more elaborate construction, one can describe an immersion of $P^2(\mathbf{R})$ into $\mathbf{R}^3$ (the so-called Boy's surface which, of course, has self-intersections). For more details, see Gerd Fischer, [FISC], Commentary, Chapter 6 (by U. Pinkall).

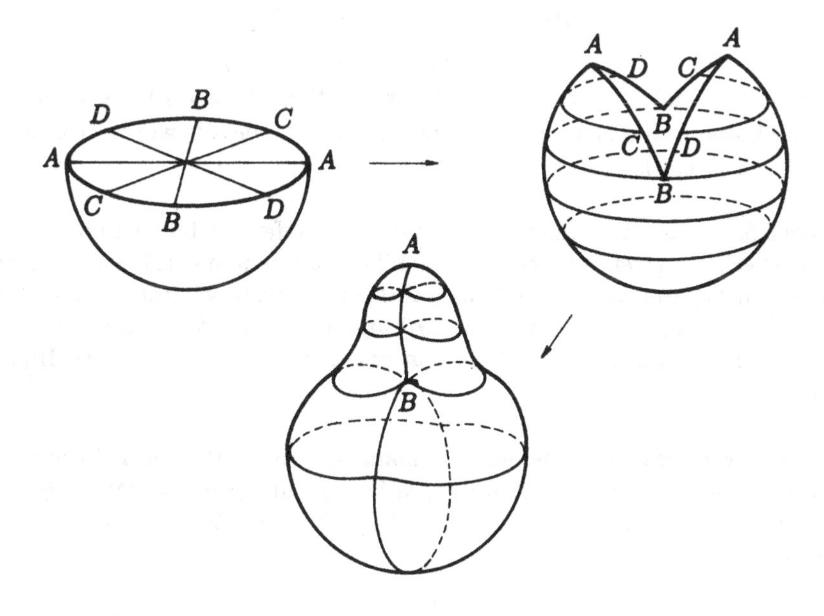

Fig. 3.6

Example 8. (Immersion of the projective plane in $\mathbf{R}^4$). Let $\varphi: \mathbf{R}^3 \to \mathbf{R}^4$ be the map given by

$$\varphi(x,y,z) = (x^2 - y^2, xy, xz, yz), \quad (x,y,z) \in \mathbf{R}^3.$$

Let $S^2 \subset \mathbf{R}^3$ be the unit sphere and let $\pi: S^2 \to P^2(\mathbf{R})$ be the canonical projection of S^2 onto the real projective plane (see Example 7). Notice that $\varphi(p) = \varphi(-p)$, and define a map $\theta: P^2(\mathbf{R}) \to \mathbf{R}^4$ by

$$\theta(\{p, -p\}) = \varphi(p).$$

Since π is a local diffeomorphism, to show that θ is an immersion, it suffices to show that the restriction of φ to the sphere S^2 is an immersion. To see that, choose the parametrization of S^2 given in Example 7. Then f_3^+, for instance, is given by

$$f_3^+(x,y) = (x, y, \sqrt{1 - (x^2 + y^2)})$$

and

$$\varphi \circ f_3^+(x,y,z) = (x^2 - y^2, xy, xD, yD).$$

To show that $d(\varphi \circ f_3^+)$ is injective, it suffices to show that the rank of the matrix

$$\begin{pmatrix} 2x & y & D + xD_x & yD_x \\ -2y & x & xD_y & D + yD_y \end{pmatrix}$$

is equal to two, and this is easily checked. Similarly, we can check that the same holds for all other parametrizations and this completes the example.

The immersion θ is actually an embedding. The details will be left as an exercise (Exerc. 5).

Remark 6. A natural question in the theory of differentiable manifolds is to know whether a given differentiable manifold can be immersed or embedded in some euclidean space. A fundamental theorem due to Whitney states that: *every differentiable manifold (Hausdorff and with countable basis) of dimension n can be immersed in $\mathbf{R}^{2n}$ and embedded in $\mathbf{R}^{2n+1}$*. See, e.g., M. Hirsh, [HIR].

We now extend for differentiable manifolds the notion of a differential form of Chapter 1. Given a vector space V, we will denote by $\Lambda^k(V)$ the set of alternate, k-linear maps $w: V \times \ldots \times V \to \mathbf{R}$, where $V \times \ldots \times V$ contains k factors.

Definition 7. Let M^n be a differentiable manifold. *An exterior k-form w in M* is the choice, for every $p \in M$, of an element $w(p)$ of the space $\Lambda^k(T_pM)^*$ of alternate k-linear forms of the tangent space T_pM.

Given an exterior k-form w and a parametrization $f_\alpha: U_\alpha \to M^n$, around $p \in f_\alpha(U_\alpha)$, we define the *representation* of w in this parametrization as the exterior k-form w_α in $U_\alpha \subset \mathbf{R}^n$ given by

$$w_\alpha(v_1,\ldots,v_k) = w(df_\alpha(v_1),\ldots,df_\alpha(v_k)), \quad v_1,\ldots,v_k \in \mathbf{R}^n.$$

If we change coordinates to $f_\beta: U_\beta \to M^n$, $p \in f_\beta(U_\beta)$, we obtain that

$$\begin{aligned}(f_\beta^{-1}\circ f_\alpha)^* w_\beta(v_1,\ldots,v_k) &= w_\beta(d(f_\beta^{-1}\circ f_\alpha)(v_1),\ldots,d(f_\beta^{-1}\circ f_\alpha)(v_k))\\ &= w((df_\beta \circ d(f_\beta^{-1}\circ f_\alpha))(v_1),\ldots,(df_\beta\circ d(f_\beta^{-1}\circ f_\alpha))(v_k)\\ &= w_\alpha(v_1,\ldots,v_k),\end{aligned}$$

that is, $(f_\beta^{-1}\circ f_\alpha)^* w_\beta = w_\alpha$.

Definition 8. A *differential form of order k* (or a differential k-form) in a differentiable manifold M^n is an exterior k-form such that, in some coordinate system (hence, in all), its representation is differentiable.

From the above, it follows that a differential k-form in M^n is the choice, for each parametrization (U_α, f_α) of M, of a differential k-form w_α in U_α in such a way that for another parametrization (U_β, f_β), with $f_\alpha(U_\alpha)\cap f_\beta(U_\beta) \neq \phi$, we have $w_\alpha = (f_\beta^{-1}\circ f_\alpha)^* w_\beta$.

It is an important fact that all the operations defined for differential forms in $\mathbf{R}^n$ can be extend to differential forms in M^n through their local representations. For instance, if w is a differential form in M, dw is the differential form in M whose local representation is dw_α. Since

$$dw_\alpha = d(f_\beta^{-1}\circ f_\alpha)^* w_\beta = (f_\beta^{-1}\circ f_\alpha)^* dw_\beta,$$

dw is a well defined differential form on M.

Closely associated with differential forms is the notion of a vector field.

Definition 9. A *vector field* X on a differentiable manifold M is a correspondence that associates to each point $p \in M$ a vector $X(p) \in T_pM$. The vector field X is *differentiable* if for every differentiable function $\varphi: M \to \mathbf{R}$, $X\varphi$ is again a differentiable function.

Let $f_\alpha: U_\alpha \to M^n$ be a parametrization of M and $X_i = \frac{\partial}{\partial x_i}$, $i = 1,\ldots,n$, the basis associated to the parametrization. Then a vector field X can be written in $f_\alpha(U_\alpha)$ as

$$X = \sum_i a_i X_i.$$

Since a vector field X on M is an operation on the space D of differentiable functions of M, we can take the iterates of this operation. For instance, if X and Y are differentiable vector fields and $\varphi: M \to \mathbf{R}$ is a differentiable function, we can consider the functions $Y(X\varphi)$ and $X(Y\varphi)$. In general, such iterated operations do not lead to vector fields, since they involve derivatives of order higher than the first. However the following holds.

Lemma 2. *Let X and Y be differentiable vector fields on a differentiable manifold M. Then there exists a unique vector field Z on M such that, for each $\varphi \in D$, $Z\varphi = (XY - YX)\varphi$.*

Proof. We first prove that if such a Z exists, then it is unique. For that, let $f: U \to M$ be a parametrization, and let

$$X = \sum_i a_i \frac{\partial}{\partial x_i}, \qquad Y = \sum b_i \frac{\partial}{\partial x_i}$$

be the expressions of X and Y, respectively, in the parametrization f. Then

$$XY\varphi = X(\left(\sum_i b_j \frac{\partial \varphi}{\partial x_j}\right) = \sum_{ij} a_i \frac{\partial b_j}{\partial x_i}\frac{\partial \varphi}{\partial x_j} + \sum_{ij} a_i b_j \frac{\partial^2 \varphi}{\partial x_i \partial x_j},$$

$$YX\varphi = Y(\left(\sum_i a_i \frac{\partial \varphi}{\partial x_i}\right) = \sum_{ij} b_j \frac{\partial a_i}{\partial x_j}\frac{\partial \varphi}{\partial x_i} + \sum_{ij} a_i b_j \frac{\partial^2 \varphi}{\partial x_i \partial x_j},$$

hence

$$(XY - YX)\varphi = \sum_j \left(\sum_i \left(a_i \frac{\partial b_j}{\partial x_i} - b_i \frac{\partial a_j}{\partial x_i}\right) \frac{\partial}{\partial x_j}\right) \varphi.$$

It follows that if a Z exists with the required property, it must be expressed as above in any coordinate system, hence it is unique

To prove existence, just define Z_α in each coordinate neighborhood $f_\alpha(U_\alpha) \subset M$ by the above expression. By uniqueness, $Z_\alpha = Z_\beta$ in $f_\alpha(U_\alpha) \cap f_\beta(U_\beta)$, hence Z is well defined on M. □

Definition 10. The vector field determined by the above lemma is called the *bracket* $[X, Y] = XY - YX$ of X and Y, and it is clearly differentiable.

The bracket operation has the following properties:

Proposition 1. *Let X, Y and Z be differentiable vector fields, a and b be real numbers, and φ and θ be differentiable functions. Then*
a) $[X, Y] = -[Y, X]$,
b) $[aX + bY, Z] = a[X, Z] + b[Y, Z]$,
c) $[[X, Y], Z] + [[Y, Z], X] + [[Z, X], Y] = 0$ *(Jacobi's identity),*
d) $[\theta X, \varphi Y] = \theta\varphi[X, Y] + \theta \cdot X(\varphi)Y - \varphi \cdot Y(\theta)X$.

Proof. (a) and (b) are immediate. To prove (c), we observe that

$$\begin{aligned}[[X, Y], Z] &= [XY - YX, Z] = XYZ - YXZ - ZXY + ZYX \\ &= [X, [Y, Z]] + [Y, [Z, X]],\end{aligned}$$

and use (a) to obtain (c). The proof of (d) is a direct and simple computation. □

There exists an interesting relation between exterior differentiation of differential forms and the bracket operation. For the case of 1-forms, this relation is as follows.

Proposition 2. *Let ω be a differentiable 1-form on a differentiable manifold M and let X and Y be differentiable vector fields on M. Then*

$$d\omega(X,Y) = X\omega(Y) - Y\omega(X) - \omega([X,Y]) \tag{2}$$

Proof. Set $f: U \to M$ be a parametrization of M and let

$$X = \sum_j a_i \frac{\partial}{\partial x_i}, \quad Y = \sum_j b_j \frac{\partial}{\partial x_j}$$

be the expressions of X and Y in this parametrization. We first observe that if (2) holds for X_i and Y_j, then it also holds for $\Sigma_i X_i$ and $\Sigma_j Y_j$. Next, we claim that if (2) holds for X and Y, it also holds for θX and φY, when θ and φ are differentiable functions.

To see that, we first notice that, by hypothesis,

$$d\omega(\theta X, \varphi Y) = \theta\varphi d\omega(X,Y) = \theta\varphi\{X\omega(Y) - Y\omega(X) - \omega([X,Y])\}.$$

By using (d) of Proposition 1, we obtain

$$\begin{aligned}
(\theta X)\omega(\varphi Y) &- (\varphi Y)\omega(\theta X) - \omega([\theta X, \varphi Y]) \\
&= \theta X(\varphi)\omega(Y) + (\theta\varphi X)\omega(Y) - \varphi Y(\theta)\omega(X) - (\varphi\theta Y)\omega(X) \\
&\quad - \theta\varphi\omega([X,Y]) - \theta X(\varphi)\omega(Y) + \varphi Y(\theta)\omega(X) \\
&= \theta\varphi\{X\omega(Y) - Y\omega(X) - \omega([X,Y])\} \\
&= d\omega(\theta X, \varphi Y)
\end{aligned}$$

and this proves our claim.

It follows that it suffices to prove (2) for the vectors $\frac{\partial}{\partial x_i}, \frac{\partial}{\partial x_j}$. Since $\left[\frac{\partial}{\partial x_i}, \frac{\partial}{\partial x_j}\right] = 0$, it suffices to prove that

$$d\omega\left(\frac{\partial}{\partial x_i}, \frac{\partial}{\partial x_j}\right) = \frac{\partial}{\partial x_i}\omega\left(\frac{\partial}{\partial x_j}\right) - \frac{\partial}{\partial x_j}\omega\left(\frac{\partial}{\partial x_i}\right). \tag{2'}$$

Notice that if (2′) holds for ω_1 and ω_2, it also holds for $\omega_1 + \omega_2$. Thus, it suffices to show that

$$d(\alpha dx_k)\left(\frac{\partial}{\partial x_i}, \frac{\partial}{\partial x_j}\right) = \frac{\partial}{\partial x_i}\alpha dx_k\left(\frac{\partial}{\partial x_j}\right) - \frac{\partial}{\partial x_j}\alpha dx_k\left(\frac{\partial}{\partial x_i}\right),$$

where α is a differentiable function. The above reduces to

$$(d\alpha \wedge dx_k)\left(\frac{\partial}{\partial x_i}, \frac{\partial}{\partial x_j}\right) = \delta_{kj}\frac{\partial \alpha}{\partial x_i} - \delta_{ki}\frac{\partial \alpha}{\partial x_j}$$

which holds by the very definition of exterior product. □

Remark 7. With essentially the same proof, one can show the following generalization of Proposition 2. Let ω be a differentiable k-form and $X_1, \cdots, X_{k+1}$ differentiable vector fields. Then

$$\begin{aligned} d\omega(X_1, \cdots, X_{k+1}) = & \sum_{i=1}^{k+1} (-1)^{i+1} X_i \omega(X_1, \cdots, \hat{X}_i, \cdots, X_{k+1}) \\ & + \sum_{i<j} (-1)^{i+j} \omega([X_i, X_j], X_1, \cdots, \hat{X}_i, \cdots, \hat{X}_j, \cdots, X_{k+1}) \end{aligned}$$

where $\hat{X}_i$ means that X_i is missing.

We will conclude this chapter with the global notion of orientability for manifolds.

Definition 11. A differentiable manifold M is *orientable* if M has a differentiable structure $\{(U_\alpha, f_\alpha)\}$ such that for each pair α, β with $f_\alpha(U_\alpha) \cap f_\beta(U_\beta) \neq \phi$, the differential of the change of coordinates $f_\beta^{-1} \circ f_\alpha$ has positive determinant. Otherwise, M is called *nonorientable*.

If M is orientable, the choice of a differentiable structure satisfying the above is called an *orientation* for M.

Examples of orientable and nonorientable manifolds are given in the Exercises.

Remark 8. It can be shown that a compact regular surface in $\mathbf{R}^3$ is orientable. A beautiful and simple proof of this fact can be found in Elon Lima [LIM 2]. As we will see in Exercises 10 and 11, the projective plane and the Klein bottle are not orientable; hence, they cannot be embedded in $\mathbf{R}^3$, a fact mentioned earlier in Remarks 4 and 5 .

EXERCISES

1) Show with details that the real projective space $P^n(\mathbf{R})$ is a differentiable manifold.

2) Let M and N be differentiable manifolds where $\{(U_\alpha, f_\alpha)\}$ is a differentiable structure for M and $\{(V_\beta, g_\beta)\}$ is a differentiable structure for N. Consider the cartesian product $M \times N$ and the maps $h_{\alpha\beta}: U_\alpha \times V_\beta \to M \times N$ given by

$$h_{\alpha\beta}(x, y) = (f_\alpha(x), g_\beta(y)), \quad x \in U_\alpha, \quad y \in V_\beta.$$

Show that $\{(U_\alpha \times V_\beta, h_{\alpha\beta})\}$ is a differentiable structure for $M \times N$ which is then called the *product manifold* of M and N. Describe the product manifold $S^1 \times S^1$ of two circles, where S^1 has the usual differentiable structure.

3) Let $\varphi: M \to N$ be a differentiable map. Show that the definition of the differential $d\varphi_p: T_pM \to T_{\varphi(p)}N$ of φ at p (Definition 4) does not depend on the choice of the curve and that $d\varphi_p$ is a linear map.

4) Let $\varphi: M \to N$ be an immersion and let p be a point in M. Show that there exists a neighborhood $V \subset M$ of p such that the restriction $\varphi|V$ of φ to V is an embedding (This means that every immersion is *locally* an embedding).

5) Prove that the immersion of $P^2(\mathbf{R})$ into $\mathbf{R}^4$ given in Example 8 is an embedding.

   Hint. To show the injectivity of θ set

$$x^2 - y^2 = a, \quad xy = b, \quad xz = c, \quad yz = d. \qquad (*)$$

   It suffices to check that, under the condition $x^2 + y^2 + z^2 = 1$, the above equations have only two solutions which are of the form (x, y, z) and $(-x, -y, -z)$. The three last equations give:

$$x^2 d = bc, \quad y^2 c = bd, \quad z^2 b = cd. \qquad (**)$$

   If b, c, d are all zero, the equation (*) shows that at least two of the coordinates x, y, z are zero, the remaining one being ± 1, since $x^2 + y^2 + z^2 = 1$. If one of the values b, c, d is non zero the equation (**) together with $x^2+y^2+z^2 = 1$ will determine x^2, y^2, z^2. From equations (*), we see that the choice of a sign for one of the x, y, z determines the signs of the remaining two.

6) Consider the cylinder $C = \{(x, y, z) \in \mathbf{R}^3; x^2 + y^2 = 1\}$ and identify the point (x, y, z) with $(-x, -y, -z)$. Show that the quotient space of C by this equivalence relation can be given a differentiable structure (infinite Möbius band).

7) Show that the tangent bundle of a differentiable manifold is orientable (even if the manifold is not so).

8) Let M be a differentiable manifold that can be covered by two coordinate neighborhoods V_1 and V_2 in such a way that the intersection $V_1 \cap V_2$ is connected. Show that M is orientable.

9) Show that the sphere $S^n = \{p \in \mathbf{R}^{n+1}; |p| = 1\}$ is orientable.

10) Show that the real projective plane $P^2(\mathbf{R})$ is not orientable.
Hint. Show that if a manifold M is orientable, any open set in M is an orientable manifold. Notice that $P^2(\mathbf{R})$ contains an open Möbius band that is not orientable (cf. [dC], §2.6, Example 3).
11) Show that the Klein bottle is not orientable.
12) A *field of planes* in an open set $U \subset \mathbf{R}^3$ is a correspondence P that associates to each $p \in U$ a plane $P(p)$ passing through p. The field P is differentiable if the coefficients of the equation of $P(p)$ are differentiable functions in p. A *integral surface* of P is a surface $S \subset \mathbf{R}^3$ such that for each $q \in S$, we have $T_qS = P(q)$, i.e., S is tangent at each of its points to the plane of the field passing there.
Let ω be a differentiable 1-form in $U \subset \mathbf{R}^3$ with $\omega(q) \neq 0, q \in U$. Show that:
(a) ω determines a differentiable plane field P by the condition

$$v \in P(p) \subset \mathbf{R}^3 \Longleftrightarrow \omega_p(v) = 0.$$

(b) If S is an integral surface of P passing through p, then $i^*(\omega) = 0$, where $i: S \subset \mathbf{R}^3$ is the inclusion.
(c) If there exists an integral surface S of P passing through p, for all $p \in U$, there exists a 1-form σ in a neighborhood $V \subset U$ of p such that $d\omega = \omega \wedge \sigma$.
Hint for (c): Consider two 1-forms ω_2, ω_3 in such that $\omega = \omega_1, \omega_2, \omega_3$ are linearly independent in V, and write

$$d\omega = \alpha\omega_2 \wedge \omega_3 + \beta\omega_3 \wedge \omega_1 + \gamma\omega_1 \wedge \omega_2$$

By using the fact that $d(i^*\omega) = i^*d\omega = 0$ and that the 2-form $\omega_2 \wedge \omega_3$ is nonzero, one concludes that $\alpha = 0$, hence

$$d\omega = \omega_1 \wedge (\gamma\omega_2 - \beta\omega_3).$$

(d) If there exists an integral surface of P for all $p \in U$ and $\omega = adx + bdy + cdz$, then

$$\left(\frac{\partial c}{\partial y} - \frac{\partial b}{\partial z}\right) a + \left(\frac{\partial a}{\partial z} - \frac{\partial c}{\partial x}\right) b + \left(\frac{\partial b}{\partial x} - \frac{\partial a}{\partial y}\right) c = 0$$

Hint for (d): One concludes from (c) that $d\omega \wedge \omega = 0$ and writes this equation in coordinates.
13) Set $\omega = xdx + ydy + zdz$, and let P be the field of planes in $\mathbf{R}^3 - \{0\}$ determined by ω. Show that the integral surface of P passing through $p = (x, y, z)$ is the sphere with center in the origin (0,0,0) and passing through p.
14) Set $\omega = zdx + xdy + ydz$. Show that the plane field determined by ω has no integral surface.

15) Consider in the real line $\mathbf{R}$ the two following differentiable structures: 1) $(\mathbf{R}, f_1)$ where $f_1(x) = x$. 2) $(\mathbf{R}, f_2)$, where $f_2(x) = x^3$. Show that:
    a) The identity map $i: (\mathbf{R}, f_1) \to (\mathbf{R}, f_2)$, $i(x) = x$, is not a diffeomorphism (thus the maximal structures determined by $(\mathbf{R}, f_1)$ and $(\mathbf{R}, f_2)$ are distinct)
    b) The map $\varphi: (\mathbf{R}, f_1) \to (\mathbf{R}, f_2)$ given by $\varphi(x) = x^3$ is a diffeomorphism (thus, although the differentiable structures are distinct, they define differentiable manifolds that are diffeomorphic).
16) (*The orientable double covering*). Let M be a connected differentiable manifold. For each $p \in M$, denote by $\mathcal{O}_p$ the quotient space of the set of all bases of T_pM under the following equivalence relation: two bases are equivalent if they are related by a matrix with positive determinant. Clearly $\mathcal{O}_p$ has two elements, and each element O_p of $\mathcal{O}_p$ is called an *orientation* at p. Now let

$$\tilde{M} = \{(p, O_p); p \in M, O_p \in \mathcal{O}_p\},$$

and let $f_\alpha: U_\alpha \to M$ be a parametrization of M with $p \in f_\alpha(U_\alpha)$. Define $\tilde{f}_\alpha: U_\alpha \to \tilde{M}$ by

$$\tilde{f}_\alpha(x_1, \ldots, x_n) = (f_\alpha(x_1, \ldots, x_n), [\frac{\partial}{\partial x_1}, \ldots, \frac{\partial}{\partial x_n}])$$

where $(x_1, \ldots, x_n) \in U_\alpha$, and $[\frac{\partial}{\partial x_1}, \ldots, \frac{\partial}{\partial x_n}]$ denotes the element of $\mathcal{O}_p$ determined by this basis. Show that:
    a) If $\{(U_\alpha, f_\alpha)\}$ is a differentiable structure in M, then $\{(U_\alpha, \tilde{f}_\alpha)\}$ is a differentiable structure in $\tilde{M}$ which is orientable (even if M is not).
    b) The map $\pi: \tilde{M} \to M$ given by $\pi(p, O_p) = p$ is differentiable, surjective, and each point $p \in M$ has a neighborhood V whose inverse image $\pi^{-1}(V)$ is the disjoint union of two open sets each of which is applied by π diffeomorphically onto V. By this reason, $\tilde{M}$ is called the *orientable double covering* of M.
    c) M is orientable if and only if $\tilde{M}$ is not connected.

    Hint. Notice that if M is orientable, the sets $M_1 = \{(p$, orient. of M in $p)\}$, $M_2 = \{(p$, orient. opposite to that of M in $p)\}$, are nonempty, disjoint, open subsets of $\tilde{M}$. For the converse, one has to show first that the image $\pi(F)$ of a closed set $F \subset \tilde{M}$ is a closed set in M; this follows from the fact that for each $p \in M$, the inverse image $\pi^{-1}(p)$ has two points. Now, if $\tilde{M}$ is not connected, denote by C a connected component of $\tilde{M}$. Then $\pi(C)$ is an open, closed and nonvoid subset of M, hence $\pi(C) = M$. Therefore, $\tilde{M}$ is the disjoint union of two connected components and π applies diffeomorphically each such component onto M. It follows that M is orientable.
17) (*A non-Hausdorff manifold*). Let S be the set given by the disjoint union of $\mathbf{R}^2$ with a point p^*. Let f_1 and f_2 maps of $\mathbf{R}^2$ in S defined by

$$f_1(u,v) = f_2(u,v), \text{if } (u,v) \neq (0,0), (u,v) \in \mathbf{R}^2,$$
$$f_1(0,0) = (0,0),$$
$$f_2(0,0) = p^*.$$

Show that $(\mathbf{R}^2, f_i)$, $i = 1, 2$, is a differentiable structure in S the topology of which does not satisfy the axiom of Hausdorff.

4. Integration on Manifolds; Stokes Theorem and Poincaré's Lemma

1. Integration of Differential Forms

In this section we will define the integral of a differential n-form on an n-dimensional differentiable manifold. We will start with the case of $\mathbf{R}^n$.

Let ω be a differential form defined in an open set $U \subset M^n$. The *support* K of ω is the closure of the set

$$A = \{p \in M^n; \quad \omega(p) \neq 0\}.$$

Let ω be a n-form and $M^n = \mathbf{R}^n$. Then

$$\omega = a(x_1, \ldots, x_n) dx_1 \wedge \ldots \wedge dx_n.$$

Assume that the support K of ω is compact and contained in U. We define

$$\int_U \omega = \int_K a\, dx_1 \ldots dx_n,$$

where the right hand side is the usual multiple integral in $\mathbf{R}^n$.

We will now proceed to the definition of the integral of an n-form on M^n. To avoid convergence problems, it is convenient to assume that M is compact; then the support K of ω, being a closed set in a compact space, is also compact. As we will see in a while, it is also necessary to assume that M is orientable, i.e., that M is covered by a family of coordinate neighborhoods $\{V_\alpha\}$ such that the coordinate changes have positive jacobians.

Let us assume initially that K is contained in some coordinate neighborhood $V_\alpha = f_\alpha(U_\alpha)$. Then, if the local representation ω_α of ω in U_α is

$$\omega_\alpha = a_\alpha(x_1, \ldots, x_n) dx_1 \wedge \ldots \wedge dx_n,$$

we define

$$\int_M \omega = \int_{V_\alpha} \omega_\alpha = \int_{U_\alpha} a_\alpha dx_1, \ldots, dx_n,$$

where the right hand side is an integral in $\mathbf{R}^n$.

It may happen that K is contained in another coordinate neighborhood $V_\beta = f_\beta(U_\beta)$ of the same family, and we must show that the above definition is independent of the choice of the coordinate neighborhood.

For that, we can assume, by contracting U_α and U_β if necessary, that $V_\alpha = V_\beta$. Let the change of coordinates

$$f = f_\alpha^{-1} \circ f_\beta : U_\beta \to U_\alpha$$

be given by

$$x_i = f_i(y_1, \dots, y_n), \quad i = 1, \dots, n,$$
$$(x_1, \dots, x_n) \in U_\alpha, \qquad (y_1, \dots, y_n) \in U_\beta.$$

Since $\omega_\beta = f^*(\omega_\alpha)$, we obtain that

$$\omega_\beta = \det(df) a_\beta dy_1 \wedge \dots \wedge dy_n,$$

where

$$a_\beta = a_\alpha(f_1(y_1, \dots, y_n), \dots, f_n(y_1, \dots, y_n)).$$

On the other hand, by the formula of change of variables for multiple integrals in $\mathbf{R}^n$, we obtain that

$$\int_{U_\alpha} a_\alpha dx_1 \dots dx_n = \int_{U_\beta} \det(df) a_\beta dy_1, \dots dy_n.$$

Thus, since $\det(df) > 0$,

$$\int_{V_\alpha} \omega_\alpha = \int_{V_\beta} \omega_\beta,$$

hence the asserted independence.

Notice that without the hypothesis of orientability for M, the sign of the integral of ω is not well defined. The choice of an orientation for M fixes a sign for the integral of ω which changes with the change of orientation.

Let us now consider the case in which the support K of ω is contained in no coordinate neighborhood. For that, we need some preliminaries, and before going into details, we will present a sketch of what we intend to do.

Given a covering $\{V_\alpha\}$ of a compact differentiable manifold M, we will construct a finite family of differentiable functions $\varphi_1, \dots, \varphi_m$ such that:

a) $\sum_{i=1}^{m} \varphi_i = 1$,

b) $0 \leq \varphi_i \leq 1$, and the support of φ_i is contained in some $V_{\alpha_i} = V_i$.

The family $\{\varphi_i\}$ is called a *differentiable partition of unity* subordinate to the covering $\{V_\alpha\}$. (When M is orientable, we choose $\{V_\alpha\}$ compatible with the orientation).

Let us assume, for the time being, the existence of such a family. Assume furthermore that M is orientable. We will define the integral of an n-form ω on M^n as follows.

The support of the form $\varphi_i\omega$ is contained in V_i. By a previous definition, it makes sense to write

$$\int_M \omega = \sum_{i=1}^{m} \int_M \varphi_i \omega,$$

the only question being whether this definition is independent of the choices made.

Consider another covering $\{W_\beta\}$ of M which determines on M the same orientation as $\{V_\alpha\}$, and let $\{\psi_j\}$, $j = 1, \dots, s$, be a partition of unity subordinate to $\{W_\beta\}$. Then $\{V_\alpha \cap W_\beta\}$ will be a covering for M and the family $\varphi_i \psi_j$ will be a partition of unity subordinate to $\{V_\alpha \cap W_\beta\}$. Thus

$$\sum_{i=1}^{m} \int_M \varphi_i \omega = \sum_{i=1}^{m} \int_M \varphi_i \left(\sum_{j=1}^{s} \psi_j \right) \omega = \sum_{ij} \int_M \varphi_i \psi_j \omega,$$

where in the last equality it was used that, for each i, the functions $\varphi_i \psi_j$ are defined in V_i. Similarly,

$$\sum_{j=1}^{s} \int_M \psi_j \omega = \sum_{j=1}^{s} \int_M \left(\sum_{i=1}^{m} \varphi_i \right) \psi_j \omega = \sum_{ij} \int_M \varphi_i \psi_j \omega$$

which proves the required independence.

Remark 1. What we have done was essentially the following. We observed that the integral of a differential form whose domain is contained in a coordinate neighborhood reduces to a multiple integral. To integrate differential forms in more complicated domains we can proceed in either of the two following ways: We divide the complicated domain in simpler domains, and add up the results, or we decompose the form into forms that are zero outside simple domains and add up the results. Since it is easier to work with functions than with domains, the second alternative is usually preferred, and it was the one used here.

We now go into the proof of existence of a differentiable partition of unity subordinate to a given covering by coordinate neighborhoods of a compact differentiable manifold M^n (we will require no orientability for that).

In what follows, $B_r(0) = \{p \in \mathbf{R}^n;\ |p| < r\}$.

Lemma 1. *There exists a differentiable function* $\varphi: B_3(0) \to \mathbf{R}$ *such that:*

a) $\varphi(p) = 1$, *if* $p \in B_1(0)$

b) $0 < \varphi(p) \leq 1$, *if* $p \in B_2(0)$

c) $\varphi(p) = 0$, *if* $p \in B_3(0) - B_2(0)$.

Proof. We first consider the function $\alpha: \mathbf{R} \to \mathbf{R}$ given by (Fig. 1 (a))

$$\alpha(t) = e^{-\frac{1}{(t+1)(t+2)}}, \qquad t \in (-2, -1).$$
$$\alpha(t) = 0 \qquad , \qquad t \notin (-2, -1).$$

Notice that the function α is a simple modification of the well known function $e^{-(1/x^2)}$, and the point is that it is C^∞ everywhere.

Now take the integral

$$\gamma(t) = \int_{-\infty}^{t} \alpha(s)ds$$

to obtain a differentiable function γ (Fig. 1, (b)) whose maximum value (at $t = -1$) is given by $\int_{-2}^{-1} \alpha(s)ds = A$. Then, by setting $\beta(t) = \gamma(t)/A$, we obtain a differentiable function with the following properties:

$$\begin{aligned} \beta(t) &= 0, \quad \text{if} \quad t \leq -2, \\ 0 < \beta(t) &\leq 1, \quad \text{if} \quad t \in (-2,-1), \\ \beta(t) &= 1, \quad \text{if} \quad t \geq -1. \end{aligned}$$

The required function $\varphi: B_3(0) \to \mathbf{R}$ is obtained by $\varphi(p) = \beta(-|p|)$, $p \in B_3(0)$; for the case of $\mathbf{R}^2$ it has the form of Fig. 1 (c). □

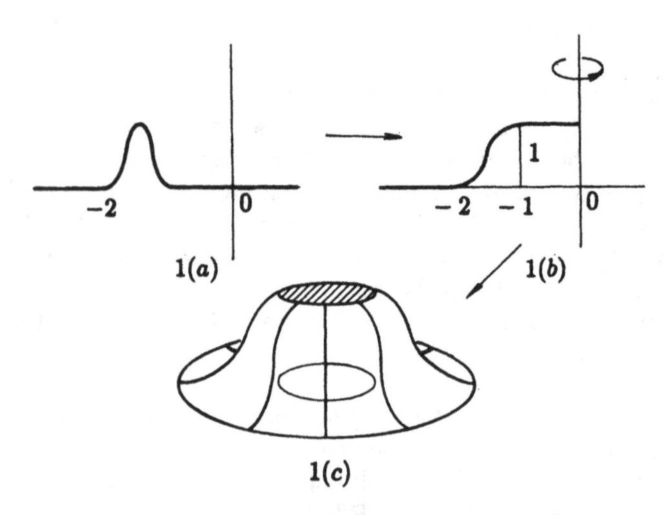

Fig. 4.1(a) **Fig. 4.1(b)** **Fig. 4.1(c)**

Lemma 2. *Let M^n be a differentiable manifold, let $p \in M$ and let $g: U \subset \mathbf{R}^n \to M$ be a parametrization around p. Then, it is possible to obtain a parametrization $f: B_3(0) \to M$ around p in such a way that $f(B_3(0)) \subset g(U)$ and that $f^{-1}(p) = (0,\ldots,0)$.*

Proof. Let $(x_1^0,\ldots,x_n^0) \in U$ be such that $g(x_1^0,\ldots,x_n^0) = p$. Since U is open, there exists an $r > 0$ such that $B_r(x_1^0,\ldots,x_n^0) \subset U$. Let T the translation in $\mathbf{R}^n$ that takes $(x_1^0,\ldots,x_n^0)$ to $(0,\ldots,0)$, and let $H: \mathbf{R}^n \to \mathbf{R}^n$ be the map that to each $p \in \mathbf{R}^n$ associates the point $\frac{3}{r}p$. Then $H \circ T$ takes $B_r(x_1^0, \dots, x_n^0)$ to $B_3(0)$.

We define the parametrization $f: B_3(0) \to M$ by

$$f = g \circ T^{-1} \circ H^{-1}$$

which is easily seen to satisfy the required conditions. □

Proposition 1. *(Existence of a differentiable partition of unity). Let M be a compact manifold and let $\{V_\alpha\}$ be a covering of M by coordinate neighborhoods. Then there exist differentiable functions $\varphi_1, \dots, \varphi_m$ such that:*

a) $\sum_{i=1}^{m} \varphi_i = 1$

b) $0 \leq \varphi_i \leq 1$, *and the support of φ_i is contained in some V_{α_i} of the covering $\{V_\alpha\}$.*

Proof. For each $p \in M$ consider the parametrization $f_p: B_3(0) \to M$ given by Lemma 2 with $f_p(B_3(0) = V_p \subset V_\alpha$, for some V_α of the covering $\{V_\alpha\}$. Set $W_p = f_p(B_1(0)) \subset V_p$.

The family $\{W_p\}$ is an open covering of M. Since M is compact, we can select from it a finite covering $W_1, \dots, W_m$. The corresponding $V_1, \dots, V_m$ will make up a covering of M.

Let us define functions $\theta_i: M \to \mathbf{R}$, $\quad i = 1, \dots, m$, by

$$\theta_i = \varphi \circ f_i^{-1} \text{ in } V_i; \quad \theta_i = 0 \text{ in } M - V_i,$$

where $\varphi : B_3(0) \to \mathbf{R}$ is the function given by Lemma 1. The functions θ_i are differentiable and the support of θ_i is contained in V_i.

Finally define φ_i by

$$\varphi_i(p) = \frac{\theta_i(p)}{\sum_{j=1}^{m} \theta_j(p)}, \quad p \in M.$$

It is immediate to check that the functions φ_i so constructed satisfy the conditions (a) and (b). □

Remark 2. The existence of a differentiable partition of unity is one of the most useful facts for the study of global questions on differentiable manifolds. Proposition 1 still holds for noncompact manifolds (with countable basis) by considering locally finite countable coverings (locally finite means that each point of the manifold meets only a finite number of members of the covering), and the proof is essentially the same. That every manifold has a locally finite countable covering can be found, for instance, in F. Warner, [WAR].

2. Stokes Theorem

In this section we intend to establish Stokes theorem. For that, we will need a series of definitions that will make it possible to state precisely the theorem; once stated, the proof of the theorem is relatively simple.

For the two-dimensional case, a rough description of the theorem is as follows.

Let ω be a differential 1-form defined on a two-dimensional, oriented, manifold M^2, and let $d\omega$ be its exterior differential. Consider a region $\mathbf{R}$ of M^2 *bounded* by a closed regular curve $C = \partial R$. The orientation of $\mathbf{R}$ induces an orientation on C, and the inclusion $i: C \to M$ allows us to consider the *restriction* $i^*\omega$ of ω to C. Under these conditions, Stokes theorem states that the integral of the 2-form $d\omega$ in $\mathbf{R}$ is equal to the integral of $i^*\omega$ in $\partial R = C$. So, in a certain sense, the *operators* d (applied to forms) and ∂ (applied to smooth domains) are dual to each other.

If, in particular, $M^2 = \mathbf{R}^2$ and $\omega = Pdx + Qdy$, Stokes theorem reduces to Gauss theorem:

$$\int\int_R \left(\frac{\partial Q}{\partial x} - \frac{\partial P}{\partial y}\right) dxdy = \int_C Pdx + Qdy.$$

We now start to present the definitions we need and which are useful in other contexts. The first one is an extension of the notion of a manifold to include manifolds with "boundary". The definition of a manifold does not include, for instance, the set M, given by

$$M = \{(x, y, z) \in \mathbf{R}^3; \quad z = x^2 + y^2, \quad z \leq z_0, \quad z_0 > 0\}$$

(M is the closed set of the rotation paraboloid bounded above by $z = z_0$), because the intersection $V \cap M$ of any neighborhood V of a point $p = (x, y, z_0)$ in the "boundary" of M with M is not homeomorphic to an open set of $\mathbf{R}^2$ (Fig. 4.2). Notice, however, that $V \cap M$ is homeomorphic to an open set of the closed half-space $\{(x_1, x_2) \in \mathbf{R}^2; \ x_1 \leq 0\}$, whereas points of M that are not in the boundary behave as points in a 2-manifold. This suggests a new definition that will include the above situation.

A *half-space of* $\mathbf{R}^n$ is the set

$$H^n = \{(x_1, \dots, x_n) \in \mathbf{R}^n; \ x_1 \leq 0\}.$$

An *open set* of H^n is the intersection with H^n of an open U of $\mathbf{R}^n$.

We say that a function $f: V \to \mathbf{R}$ defined in an open set V of H^n is *differentiable* if there exists an open set $U \supset V$ and a differentiable function $\bar{f}$ in U such that the restriction of $\bar{f}$ to V is equal to f. In this case, the differential df_p, $p \in V$, of f at p is defined to be $df_p = d\bar{f}_p$.

When V does not contain points of the form $(0, x_2, \dots, x_n)$, V is an open set of $\mathbf{R}^n$ and the definition of df_p agrees with the usual one. If p is of the form

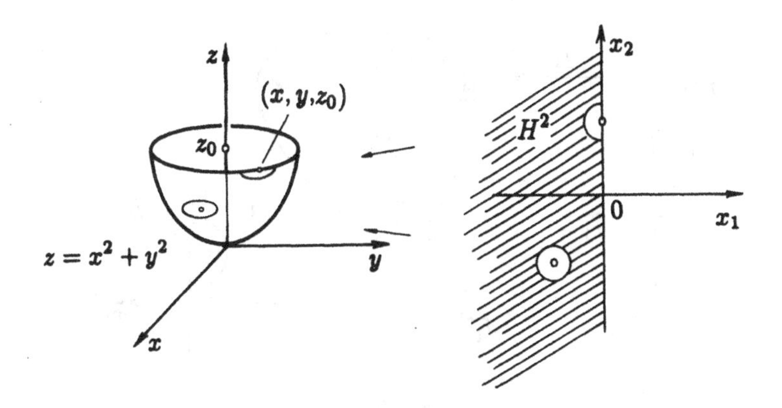

Fig. 4.2

$(0, x_2, \dots, x_n)$, df_p is defined for all tangent vector of curves in U passing through p, i.e., for *all* vectors in $\mathbf{R}^n$ with origin in p. Using such curves it is easy to show that the definition of df_p is independent of the extension $\overline{f}$ of f.

In a similar way, we define a differentiable map $f: V \to \mathbf{R}^n$.

Definition 1. An *n-dimensional differentiable manifold with* (regular) *boundary* is a set M and a family of injective maps $f_\alpha: U_\alpha \subset H^n \to M$ of open sets of H^n into M such that:

1) $\bigcup_\alpha f_\alpha(U_\alpha) = M$.
2) For all pairs α, β with $f_\alpha(U_\alpha) \cap f_\beta(U_\beta) = W \neq \phi$ the sets $f_\alpha^{-1}(W)$ and $f_\beta^{-1}(W)$ are open sets in H^n and the maps $f_\beta^{-1} \circ f_\alpha$, $f_\alpha^{-1} \circ f_\beta$ are differentiable.
3) The family $\{(U_\alpha, f_\alpha)\}$ is maximal relative to (1) and (2).

A point $p \in M$ is said to be a *point in the boundary* of M if for some parametrization $f: U \subset H^n \to M$ around p we have that $f(0, x_2, \dots, x_n) = p$.

Lemma 3. *The definition of point in the boundary does not depend on parametrizations.*

Proof. Let $f_1: U_1 \to M$ be a parametrization around p such that $f_1(q) = p$, $q = (0, x_2, \dots, x_n)$.

Assume, by contradiction, that for some parametrization $f_2: U_2 \to M$ around p we have $f_2^{-1}(p) = q_2 = (x_1, \dots, x_n)$ with $x_1 \neq 0$. (Fig. 4.3)

Let $W = f_1(U_1) \cap f_2(U_2)$. The map

$$f_1^{-1} \circ f_2: f_2^{-1}(W) \to f_1^{-1}(W)$$

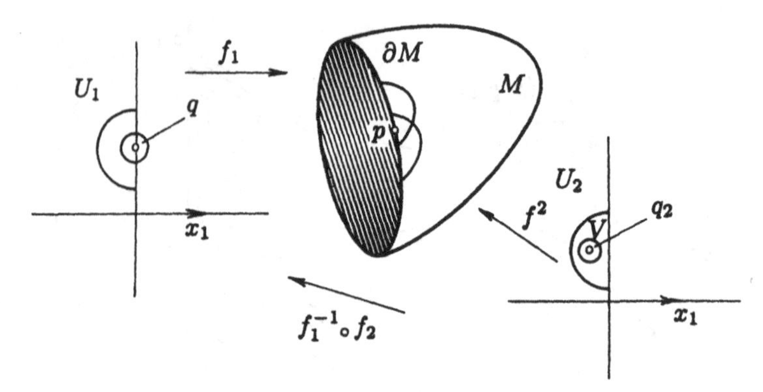

Fig. 4.3

is a diffeomorphism. Since $x_1 \neq 0$, there exists a neighborhood U of q_2, $U \subset f_2^{-1}(W)$, that does not intersect the x_1-axis. Restricting $f_1^{-1} \circ f_2$ to U, we will have a differentiable map

$$f_1^{-1} \circ f_2 \colon U \to H^n$$

such that the determinant of $d(f^{-1} \circ f_2)_{q_2}$ is nonzero. By the inverse function theorem, $f_1^{-1} \circ f_2$ will take a neighborhood $V \subset U$ of q_2 diffeomorphically onto $f_1^{-1} \circ f_2(V)$. But then $f_1^{-1} \circ f_2(V)$ would contain points of the form $(x_1, \ldots, x_n)$ with $x_1 > 0$ which are not in H^n. This yields a contradiction and completes the proof. □

The set of points in the boundary of M is therefore well defined; it is called *the boundary of* M and denoted by ∂M. If $\partial M = \phi$, Definition 1 agrees with the definition of a differentiable manifold given in Chapter 2. The definitions of differentiable functions, tangent space, orientability, etc, for manifolds with boundary are introduced in exactly the same way as the corresponding definitions for differentiable manifolds, with the additional care of replacing $\mathbf{R}^n$ by H^n.

Proposition 2. *The boundary ∂M of an n-dimensional differentiable manifold M with boundary is an $(n-1)$-differentiable manifold. Furthermore, if M is orientable, an orientation for M induces an orientation for ∂M.*

Proof. Let $p \in M$ be a point in the boundary of M and let $f_\alpha \colon U_\alpha \subset H^n \to M^n$ be a parametrization around p. Then $f_\alpha^{-1}(p) = q = (0, x_2, \ldots, x_n) \in U_\alpha$.

Let

$$\overline{U}_\alpha = U_\alpha \cap \{(x_1, \ldots, x_n) \in \mathbf{R}^n;\ x_1 = 0\}.$$

By identifying the set $\{(x_1, \ldots, x_n) \in \mathbf{R}^n;\ x_1 = 0\}$ with $\mathbf{R}^{n-1}$, we see that $\overline{U}_\alpha$ is an open set in $\mathbf{R}^{n-1}$. By denoting by $\overline{f}_\alpha$ the restriction of f_α to $\overline{U}_\alpha$, we see, by Lemma 3, that $\overline{f}_\alpha(\overline{U}_\alpha) \subset \partial M$. Finally, by letting p run in the points of ∂M, we easily check that the family $\{(\overline{U}_\alpha, \overline{f}_\alpha)\}$ is a differentiable structure for ∂M. This proves the first part of the Proposition.

To prove the second part, assume that M is orientable and choose an orientation for M, i.e., a differentiable structure $\{(U_\alpha, f_\alpha)\}$ such that the changes of coordinates have positive jacobian. Consider the elements of the family that satisfy the condition $f_\alpha(U_\alpha) \cap \partial M \neq \phi$. Then the family $\{(\overline{U}_\alpha, \overline{f}_\alpha)\}$ described in the first part is a differentiable structure for ∂M. We want to show that if $\overline{f}_\alpha(\overline{U}_\alpha) \cap \overline{f}_\beta(\overline{U}_\beta) \neq \phi$, the change of coordinates has positive jacobian, i.e., that

$$\det(d(\overline{f}_\alpha^{-1} \circ \overline{f}_\beta)_q) > 0,$$

for all q whose image, by some parametrization, is in the boundary.

Observe that the change of coordinates $f_\alpha \circ f_\beta^{-1}$ takes a point of the form $(0, x_2^\beta, \ldots, x_n^\beta)$ into a point of the form $(0, x_2^\alpha, \ldots, x_n^\alpha)$. Thus, for a point q whose image is in the boundary,

$$\det(d(f_\alpha^{-1} \circ f_\beta) = \frac{\partial x_1^\alpha}{\partial x_1^\beta} \det(d(\overline{f}_\alpha^{-1} \circ \overline{f}_\beta)).$$

But $\frac{\partial x_1^\alpha}{\partial x_1^\beta} > 0$, because $x_1^\alpha = 0$ in $q = (0, x_2^\alpha, \ldots, x_n^\alpha)$, and both x_1^α and x_1^β are negative in a neighborhood of p. Since $\det(d(f_\alpha^{-1} \circ f_\beta)) > 0$, by hypothesis, we conclude that $\det(d(\overline{f}_\alpha^{-1} \circ \overline{f}_\beta)) > 0$, as we wished. □

We can now state and prove Stokes theorem.

Theorem 1. *Let M^n be a differentiable manifold with boundary, compact and oriented. Let ω be a differential $(n-1)$-form on M, and let $i: \partial M \to M$ be the inclusion map of the boundary ∂M into M. Then*

$$\int_{\partial M} i^*\omega = \int_M d\omega.$$

Proof. Let K be the support of ω. We will consider the following cases:

A) K is contained in some coordinate neighborhood $V = f(U)$ of a parametrization $f: U \subset H^n \to M$. In U,

$$\omega = \sum_{j=1}^n a_j dx_1 \wedge \ldots \wedge dx_{j-1} \wedge dx_{j+1} \wedge \ldots \wedge dx_n,$$

where $a_j = a_j(x_1, \ldots, x_n)$ is a differentiable function on U. Thus

$$d\omega = \left(\sum_{j=1}^{n} (-1)^{j-1} \frac{\partial a_j}{\partial x_j} \right) dx_1 \wedge \ldots \wedge dx_n.$$

A₁) Assume first that $f(U) \cap \partial M = \phi$. Then ω is zero in ∂M and $i^*\omega = 0$. Thus

$$\int_{\partial M} i^*\omega = 0.$$

We will show that

$$\int_M d\omega = \int_U \left(\sum_j (-1)^{j-1} \frac{\partial a_j}{\partial x_j} \right) dx_1 \ldots dx_n = 0.$$

For that, extend the functions a_j to H^n by setting

$$\begin{aligned} a_j(x_1, \ldots, x_n) &= a_j(x_1, \ldots, x_n), \quad &\text{if} \quad (x_1, \ldots, x_n) \in U \\ a_j(x_1, \ldots, x_n) &= 0, \quad &\text{if} \quad (x_1, \ldots, x_n) \in H^n - U \end{aligned}$$

Since $f^{-1}(K) \subset U$, the functions a_j so defined are differentiable in H^n. Now let $Q \subset H^n$ be a parallelepiped given by $x_j^1 \leq x_j \leq x_j^0$, $j = 1, \ldots, n$, and containing $f^{-1}(K)$ in its interior (Fig. 4.4).

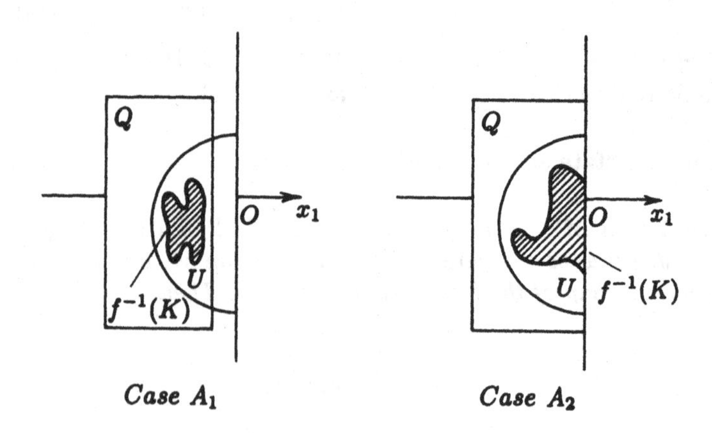

Fig. 4.4

Then

$$\int_U \left(\sum_j (-1))^{j-1} \frac{\partial a_j}{\partial x_j} \right) dx_1 \dots dx_n = \sum_j (-1)^{j-1} \int_Q \frac{\partial a_j}{\partial x_j} dx_1 \dots dx_n$$

$$= \sum_j (-1)^{j-1} \int [a_j(x_1, \dots, x_{j-1}, x_j^0, x_{j+1}, \dots x_n)$$

$$- a_j(x_1, \dots, x_{j-1}, x_j^1, x_{j+1}, \dots, x_n)] dx_1 \dots dx_{j-1}\, dx_{j+1} \dots dx_n = 0,$$

since $a_j(x_1, \dots, x_j^0, \dots, x_n) = a_j(x_1, \dots, x_j^1, \dots, x_n) = 0$, for all j.

A_2) Assume now that $f(U) \cap \partial M \neq \phi$. Then the inclusion map i can be written as: $x_1 = 0$, $x_j = x_j$. Thus, using the induced orientation on the boundary,

$$i^* \omega = a_1(0, x_2, \dots, x_n) dx_2 \wedge \dots \wedge dx_n.$$

As in case (A_1), we will extend the functions a_j to H^n, and will consider the parallelepiped Q given by

$$x_1^1 \leq x_1 \leq 0, \qquad x_j^1 \leq x_j \leq x_j^0, \qquad j = 2, \dots, n$$

and such that the union of the interior of Q with the hyperplane $x_1 = 0$ contains $f^{-1}(K)$. Then

$$\int_M d\omega = \sum_{j=1}^n (-1)^{j-1} \int_Q \frac{\partial a_j}{\partial x_j} dx_1 \cdots dx_n$$

$$= \int_Q [a_1(0, x_2, \cdots, x_n) - a_1(x_1^1, x_2, \cdots, x_n)] dx_2, \cdots, dx_n$$

$$+ \sum_{j=2}^n (-1)^{j-1} \int_Q [a_j(x_1, \cdots, x_j^0, \cdots, x_n) - a_j(x_1, \cdots, x_j^1, \cdots, x_n)]$$

$$dx_1 \cdots dx_{j-1}\, dx_{j+1} \cdots dx_n.$$

Since $a_j(x_1, \dots, x_j^0, \dots, x_n) = a_j(x_1, \dots, x_j^1, \dots, x_n) = 0$, for $j = 2, \dots, n$, and $a_1(x_1^1, x_2, \dots, x_n) = 0$, we obtain

$$\int_M \omega = \int a_1(0, x_2, \dots, x_n) dx_2 \dots dx_n = \int_{\partial M} i^* \omega.$$

B) Let us now consider the general case. Let $\{V_\alpha\}$ be a covering of M by coordinate neighborhoods compatible with the orientation, and let $\varphi_1, \dots, \varphi_m$ be a differentiable partition of unity subordinate to V_α. The forms $\omega_j = \varphi_j \omega$, $j = 1, \dots, m$ satisfy the conditions of case A. Furthermore, since $\sum_j d\varphi_j = 0$, we have

$$\sum \omega_j = \omega, \quad \sum d\omega_j = d\omega.$$

Therefore,

$$\int_M d\omega = \sum_{j=1}^m \int_M d\omega_j = \sum_{j=1}^m \int_{\partial M} i^*\omega_j$$
$$= \int_{\partial M} i^* \sum_j \omega_j = \int_{\partial M} i^*\omega.$$

□

Example. Let M be a bounded region of $\mathbf{R}^3$ such that the boundary ∂M of M is a regular hypersurface of $\mathbf{R}^3$; M is then a compact 3-dimensional manifold with boundary ∂M. Let v be a differentiable vector field in $\mathbf{R}^3$, and let ω be the 1-form in $\mathbf{R}^3$ dual to v in the natural inner product of $\mathbf{R}^3$. Then (cf. Exercise 11, Chap. 1) $d(*\omega) = (\text{div } v)\nu$, where ν is the volume element of $\mathbf{R}^3$.

Now choose an orientation for $\mathbf{R}^3$ and let N be the unit normal vector of ∂M in the induced orientation. Finally, let σ be the area element of ∂M.

Consider, in a neighborhood $U \subset \mathbf{R}^3$ of $p \in M$, differentiable orthonormal fields e_1, e_2, N such that, in the points of ∂M, e_1 and e_2 are tangent to ∂M. Then

$$i^* * \omega(e_1, e_2) = \omega(N) =< v, N >,$$

i.e., $i^*(*\omega) = \langle v, N\rangle\sigma$. Thus, in this case, Stokes theorem

$$\int_M d(*\omega) = \int_{\partial M} i^*(*\omega)$$

can be written as

$$\int_M \text{div } v\, \nu = \int_{\partial M} < v, N > \sigma$$

which is the well known divergence theorem in Analysis.

Remark. The divergence theorem is a fundamental tool in Analysis (see, for instance, the beautiful account in O. Kellog, [KELL]). Although we have proved it for regular boundaries and smooth functions, it can be generalized considerably. See, for instance, the article of D. Figueiredo, [FIG].

3. Poincaré's Lemma

Let M^n be a differentiable manifold. A differential k-form ω is said to be *exact* if there exists a $(k-1)$-form β such that $d\beta = \omega$; ω is said to be *closed* if $d\omega = 0$. Since $d^2 = 0$, an exact form is closed.

The converse of the above fact does not hold in general. For instance, let $\omega = \frac{x\,dy - y\,dx}{x^2+y^2}$ be defined in $\mathbf{R}^2 - \{(0,0)\} = U$. It is easily checked that $d\omega = 0$, i.e., ω is closed, but there exists no differentiable function g in U such that $dg = \omega$; otherwise, by Stokes theorem,

$$\int_C \omega = \int_C dg = \int_{\partial C} g = 0, \qquad C = \{(x, y) \in \mathbf{R}^2;\ x^2 + y^2 = 1\},$$

and this contradicts the fact, easily computable, that $\int_c \omega = 2\pi$. It is possible, however, to show that for each $p \in U$ there exists a neighborhood $V \subset U$ of p and a differentiable function g_V in V such that $dg_V = \omega$.

In this section, we will show that the situation of this example is completely general, that is, that the condition $d\omega = 0$ is a sufficient condition for ω to be *locally* exact (Those who are familiar with the material of Chapter 2 will notice that we are generalizing Theorem 1 of that Chapter). Actually, we will prove the result in a form slightly more general which is more convenient for the applications.

Definition 2. A differentiable manifold M is *contractible*(to some point $p_0 \in M$) if there exists a differentiable map $H\colon M \times \mathbf{R} \to M$, $H(p, t) \in M, p \in M$, $t \in R$ such that

$$H(p, 1) = p, \quad H(p, 0) = p_0, \qquad \text{for all } \ p \in M.$$

It is easy to see that $\mathbf{R}^n$ is contractible to an arbitrary point $p_0 \in \mathbf{R}^n$; it suffices to define $H(p, t) = p_0 + (p - p_0)t$. The same argument shows that the ball $B_r(0) = \{p \in \mathbf{R}^n;\ |p| < r\}$ is contractible to the origin 0. It follows that any differentiable manifold is locally contractible.

Theorem 2 (Poincaré's lemma) *Let M be a contractible differentiable manifold, and let ω be a differentiable k-form in M with $d\omega = 0$. Then ω is exact, i.e., there exists a $(k-1)$-form α in M such that $d\alpha = \omega$.*

Proof. Let $\pi\colon M \times \mathbf{R} \to M$ be the projection $\pi(p, t) = p$, and let $\overline{\omega}$ be the k-form on $M \times \mathbf{R}$ given by $\overline{\omega} = H^*\omega$, where H is the map given in the definition of contractability. We will need the following lemma.

Lemma 4. *Every k-form $\overline{\omega}$ in $M \times \mathbf{R}$ can be written uniquely as*

$$\overline{\omega} = \omega_1 + dt \wedge \eta, \tag{1}$$

where ω_1 is a k-form on $M \times \mathbf{R}$ with the property that $\omega_1(v_1, \dots, v_k) = 0$, if some v_i, $i = 1, \dots, k$, belongs to the kernel of $d\pi$, and η is a $(k-1)$-form with a similar property.

Proof of Lemma 4. Let $p \in M$ and let $f : U \to M$ be a parametrization around p. Then $f(U) \times \mathbf{R}$ is a coordinate neighborhood of $M \times \mathbf{R}$, with coordinates, say, $(x_1, \dots, x_n, t)$. In $f(U) \times \mathbf{R}$, $\overline{\omega}$ can be written as

$$\begin{aligned}\overline{\omega} = & \sum_{i_1\dots i_k} a_{i_1,\dots i_k} dx_{i_1} \wedge \dots \wedge dx_{i_k} \\ & + dt \wedge \sum b_{j_1\dots j_{k-1}} dx_{j_1} \wedge \dots \wedge dx_{j_{k-1}} \\ = & \, \omega_1 \wedge dt \wedge \eta.\end{aligned} \tag{2}$$

It is clear that ω_1 and η have the required properties. Furthermore, if the decomposition (1) holds in all of M, it has to be locally of the form (2), hence it is unique. To prove existence, we define ω_1 and η on each coordinate neighborhood by (2). In the intersection of two such neighborhoods, the definitions agree by uniqueness, thus ω_1 and η can be extended to the whole M satisfying (1). This proves Lemma 4.

Now let $i_t : M \to M \times \mathbf{R}$ the map given by $i_t(p) = (p, t)$; i_t is the inclusion of M into $M \times \mathbf{R}$ at the "level" t. We will define a map I that takes k-forms of $M \times \mathbf{R}$ into $(k-1)$-forms of M as follows: If $p \in M$ and $v_1, v_2, \dots, v_k \in T_pM$, then at p,

$$(I\overline{\omega})(v_1, \dots, v_{k-1}) = \int_0^1 \{\eta(p,t)(di_t(v_1), \dots, di_t(v_{k-1}))\} dt,$$

where η is given by the decomposition $\overline{\omega} = \omega_1 + dt \wedge \eta$ of Lemma 4.

The crucial point of the theorem is contained in the following lemma.

Lemma 5. $i_1^*\overline{\omega} - i_0^*\overline{\omega} = d(I\overline{\omega}) + I(d\overline{\omega})$.

Proof of Lemma 5. Let $p \in M$. We will use the coordinate system $(x_1, \dots, x_n, t)$ introduced in Lemma 4. We first notice that the operation I is additive, i.e., $I(\omega_1 + \omega_2) = I(\omega_1) + I(\omega_2)$. It follows that it suffices to consider the following two cases: a) $\overline{\omega} = f dx_{i_1} \wedge \dots \wedge dx_{i_k}$; b) $\overline{\omega} = f dt \wedge dx_{i_1} \wedge \dots \wedge dx_{i_{k-1}}$.

Case (a). If $\overline{\omega} = f dx_{i_1} \wedge \dots \wedge dx_{i_k}$, then $d\overline{\omega} = \frac{\partial f}{\partial t} dt \wedge dx_{i_1} \wedge \dots \wedge dx_{i_k} +$ terms without dt.

Notice that in the coordinate systems $(x_1, \dots, x_n, t)$ the operation I amounts to integrate the local representations of $\overline{\omega}$ along the second factor t. Therefore,

$$\begin{aligned}I(d\overline{\omega})(p) &= \left(\int_0^1 \frac{\partial f}{\partial t} dt\right) dx_{i_1} \wedge \dots \wedge dx_{i_k} \\ &= (f(p,1) - f(p,0)) dx_{i_1} \wedge \dots \wedge dx_{i_k} \\ &= i_1^*\overline{\omega}(p) - i_0^*\overline{\omega}(p).\end{aligned}$$

Since $I\overline{\omega} = 0$, we conclude the lemma in case (a).

Case (b). If $\overline{\omega} = f dt \wedge dx_{i_1} \wedge \dots \wedge dx_{i_{k-1}}$, then $i_1^*\overline{\omega} = 0 = i_0^*\overline{\omega}$. On the other hand,

$$d\overline{\omega} = \sum_{\alpha=1}^{n} \frac{\partial f}{\partial x_\alpha} dx_\alpha \wedge dt \wedge dx_{i_1} \wedge \ldots \wedge dx_{i_{k-1}}.$$

Therefore,

$$(Id\overline{\omega})(p) = -\sum_\alpha \left(\int_0^1 \frac{\partial f}{\partial x_\alpha} dt\right) dx_\alpha \wedge dx_{i_1} \wedge \ldots \wedge dx_{i_{k-1}}$$

and

$$\begin{aligned} d(I\overline{\omega})(p) &= d\left\{\left(\int_0^1 f dt\right) dx_{i_1} \wedge \ldots \wedge dx_{i_{k-1}}\right\} \\ &= \sum_\alpha \left(\int_0^1 \frac{\partial f}{\partial x_\alpha} dt\right) dx_\alpha \wedge dx_{i_1} \wedge \ldots \wedge dx_{i_{k-1}} \end{aligned}$$

which completes the Case (b), and the proof of the lemma. □

Now we can complete the proof of the theorem (notice that so far we have not used that $d\omega = 0$). Since M is contractible,

$$H \circ i_1 = \text{ identity}, \qquad H \circ i_0 = \text{ const.} = p_0 \in M.$$

Thus

$$\begin{aligned} \omega &= (H \circ i_1)^* \omega = i_1^*(H^*\omega) = i_1^*\overline{\omega}, \\ 0 &= (H \circ i_0)^* \omega = i_0^*(H^*\omega) = i_0^*\overline{\omega}. \end{aligned}$$

Now, since $d\omega = 0$, we obtain that $d\overline{\omega} = H^* d\omega = 0$. It follows by Lemma 5 that

$$\omega = i_1^*\omega = d(I\overline{\omega}) = d\alpha,$$

where $\alpha = I\overline{\omega}$. □

EXERCISES

1) Let $f : \mathbf{R}^3 \to \mathbf{R}$ be a differentiable function and let ω be the 2-form in $\mathbf{R}^3$ given by

$$\omega = \frac{f_x dy \wedge dz + f_y dz \wedge dx + f_z dx \wedge dy}{\sqrt{f_x^2 + f_y^2 + f_z^2}}.$$

It is well known that if $a \in \mathbf{R}$ is a regular value of f (that is, for all $p \in f^{-1}(a)$, the map df_p is surjective) then

$$M^2 = \{(x, y, z) \in \mathbf{R}^3; f(x, y, z) = a\}$$

is a regular orientable surface in $\mathbf{R}^3$ (cf. [dC], §2.2). Show that the restriction of ω to M^2 is the element of area of M^2.
Hint: We want to show that if $\{v_1, v_2\}$ is a positive basis of $T_p(M)$, $p \in M$ then $\omega(v_1, v_2) =$ area of the parallelogram made up by v_1 and v_2. Choose a parametrization $g(u, v)$ of M around p, compatible with the orientation, and notice that

$$dy \wedge dz + dz \wedge dx + dx \wedge dy = (\sum_i (g_u \wedge g_v)_i) du \wedge dv,$$

where $(g_u \wedge g_v)_i$ is the i-th coordinate of the vector product $g_u \wedge g_v$ in the canonical basis of $\mathbf{R}^3$. Since $f(x, y, z) = \text{const.}$, the vector $(f_x, f_y, f_z) = A$ lies along the positive normal of M. Thus

$$\omega = \frac{\langle A, g_u \wedge g_v \rangle}{|A|} du \wedge dv = |g_u \wedge g_v| \, du \wedge dv.$$

It is now easy to check that $\omega(g_u, g_v) =$ area (g_u, g_v), hence $\omega(v_1, v_2) =$ area (v_1, v_2).

2) a) Let $\omega = xdy - ydx$ and $j \colon M \hookrightarrow \mathbf{R}^2$ the inclusion of a bounded region with regular boundary ∂M. Show that the area of M is given by $(1/2) \int_{\partial M} j^* \omega$.
b) Let $\omega = xdy \wedge dz - ydx \wedge dz + zdx \wedge dy$ and $j \colon M \subset \mathbf{R}^3$ the inclusion of a bounded region with regular boundary ∂M. Show that the volume of M is given by $(1/3) \int_{\partial M} j^* \omega$.
c) Generalize the above for $\mathbf{R}^n$.

3) Let

$$\omega = \frac{xdy \wedge dz + ydz \wedge dx + zdx \wedge dy}{(x^2 + y^2 + z^2)^{3/2}}$$

a 2-form defined in $\mathbf{R}^3 - \{0\}$, and let $M^2 \subset \mathbf{R}^3$ be an oriented surface that does not pass through the origin $0 = (0, 0, 0)$. Show that:
a) The restriction of ω to M^2 is equal to

$$\frac{\cos \theta}{r^2} \sigma ,$$

where σ is the area element of M^2, r is the distance from 0 to a point $p \in M^2$ and θ is the positive angle from Op to the unit positive normal N to M^2 at p.
Hint: Proceed is in Exercise 1. Set $p = (x, y, z)$, $r^2 = x^2 + y^2 + z^2$, $p/r = v$. One obtains:

$$\omega = \frac{1}{r^3} \langle p, g_u \wedge g_v \rangle du \wedge dv = \frac{1}{r^2} \langle v, N \rangle \, |g_u \wedge g_v| \, du \wedge dv$$
$$= \frac{\cos \theta}{r^2} \sigma.$$

b) Define the *solid angle* under which M^2 is seen from 0 as

$$\Omega = \int_{M^2} \omega.$$

(for justification sake, notice that if $p \in M^2$, $\cos\theta \neq 0$, and ΔM is a small neighborhood around p, then $(1/r^2)\cos\theta$ (area ΔM) is the area of the region of the unit sphere with center 0 that is determined by the rays that join 0 to points of ΔM; this area is usually called the solid angle under which ΔM is see from 0). Now let M be a bounded region in $\mathbf{R}^3$ with regular boundary ∂M such that $0 \notin \partial M$, and let Ω be the solid angle under which ∂M is seen from 0. Show that

$$\Omega = 0, \text{ if } 0 \notin M, \text{ and } \Omega = 4\pi, \text{ if } 0 \in M.$$

Hint: Notice that if $0 \notin M$, $d\omega = 0$, and apply Stokes theorem.

4) Let $\varphi : \mathbf{R}^3 \to \mathbf{R}$ a differentiable function, homogenous of degree k (that is, $\varphi(tx, ty, tz) = t^k \varphi(x, y, z)$). Show that:

a) If $B = \{p \in \mathbf{R}^3; |p| \leq 1\}$ is the region bounded by the unit sphere S^2, then

$$\int_B \Delta^2 \varphi \; dx \wedge dy \wedge dz = \int_{S^2} k\varphi \; \sigma,$$

where σ is the area element of S^2 and $\Delta^2 \varphi = \varphi_{xx} + \varphi_{yy} + \varphi_{zz}$ is the Laplacian of φ.

Hint: Notice that by Euler's relation for homogeneous functions (cf. Exercise 18, Chapter 1) $x\varphi_x + y\varphi_y + z\varphi_z = k\varphi$, and use the divergence theorem.

b) Let $\varphi = a_1x^4 + a_2y^4 + a_3z^4 + 3a_4x^2y^2 + 3a_5y^2z^2 + 3a_6x^2z^2$, then

$$\int_{S^2} \varphi \; \sigma = \frac{4\pi}{5} \sum_{i=1}^{6} a_i.$$

5) Let $g : \mathbf{R}^3 \to \mathbf{R}$, $f : \mathbf{R}^3 \to \mathbf{R}$ be differentiable functions, and let $M^3 \subset \mathbf{R}^3$ be a compact differentiable manifold with boundary ∂M^2. Prove that:

a) *(first Green's identity)*

$$\int_M \langle \operatorname{grad} f, \operatorname{grad} g \rangle \nu + \int_M f \Delta^2 g \;\; \nu = \int_{\partial M} f \langle \operatorname{grad} g, N \rangle \sigma,$$

where ν and σ are, respectively, the volume element of M and the area element of ∂M, and N is the unit normal of ∂M.

Hint: Set $v = f \operatorname{grad} g$ in the divergence theorem.

b) *(second Green's identity)*

$$\int_M (f\Delta^2 g - g\Delta^2 f)\nu = \int_{\partial M} (f\langle \operatorname{grad} g, N \rangle - g\langle \operatorname{grad} f, N \rangle)\sigma.$$

6) Can one find a three-dimensional orientable differentiable manifold M^3 whose boundary is the real projective plane?

7) Let ω_1 and ω_2 be differential forms on a differentiable manifold M. Assume that ω_1 and ω_2 are closed and that ω_2 is exact. Show that $\omega_1 \wedge \omega_2$ is closed and exact.

8) Let M^n be a compact orientable manifold without boundary (i.e., $\partial M = \emptyset$) and let ω be a differential $(n-1)$-form on M^n. Show that there exists a point $p \in M$ such that $d\omega(p) = 0$.

9) Show that there exists no immersion $f : S^1 \to \mathbf{R}$ of the unit circle into the real line $\mathbf{R}$.

Hint: Use Exercise 8.

10) Let $M^2 \subset \mathbf{R}^3$ be a compact, oriented, regular surface with regular boundary ∂M, and let v be a differentiable vector field in an open set of $\mathbf{R}^3$ containing M^2.

a) Show that

$$\int_{M^2} \langle \text{rot } v, N \rangle \sigma = \int_{\partial M} \langle v, \vec{t} \rangle ds$$

where N is the unit normal field, σ the area element of M^2, $\vec{t}$ the unit tangent vector to ∂M, and ds the element of arc of ∂M.

Hint: Notice that rot $v = *(d\omega)$, where ω is the 1-form dual to v in the natural inner product of $\mathbf{R}^3$ (cf. Exercise 14, Chapter 1). By choosing local orthonormal fields e_1, e_2, N such that e_1 and e_2 are tangent to M and e_1 is tangent to ∂M, we obtain

$$d\omega(e_1, e_2) = (*d\omega)(N) = \langle \text{rot } v, N \rangle,$$
$$\omega(e_1) = \langle v, e_1 \rangle = \langle v, \vec{t} \rangle,$$

that is, $d\omega = \langle \text{rot } v, N \rangle \sigma$ and $i^*\omega = \langle v, \vec{t} \rangle ds$. Now apply Stokes theorem.

b) Let $p \in \mathbf{R}^3$, $\vec{r}$ a unit vector in $\mathbf{R}^3_p$ and P the plane normal to $\vec{r}$, passing through p and whose orientation together with $\vec{r}$ gives the orientation of $\mathbf{R}^3$. Consider a disk $D \subset P$, with center p, and apply (a) to the surface made up of D and its boundary ∂D to obtain

$$\langle \text{rot } v, \vec{r} \rangle (p) = \lim_{D \to p} \frac{1}{\text{area } D} \int_{\partial D} \langle v, \vec{t} \rangle ds,$$

where the limit is taken when D runs in a family of concentric disks that approach p.

11) *(Introduction to potential theory in* $\mathbf{R}^3$*)*

A differentiable function $g : \mathbf{R}^3 \to \mathbf{R}$ is said to be *harmonic* in a subset $B \subset \mathbf{R}^3$ if $\Delta^2 g = 0$ for all $p \in B$. Let $M \subset \mathbf{R}^3$ be a bounded region with regular boundary ∂M. Prove that:

a) If g_1 and g_2 are harmonic in M and $g_1 = g_2$ in ∂M, then $g_1 = g_2$ in M.
Hint: Use Green's first identity (Exercise 5a) with $f = g = g_1 - g_2$
b) If g is harmonic in M and

$$\frac{\partial g}{\partial N} \overset{\text{def}}{=} \langle \operatorname{grad} g, N \rangle = 0$$

in ∂M, where N is the unit normal vector of ∂M, then $g =$ const. in M.
Hint: Use Green's first identity with $f = g$.
c) If g_1 and g_2 are harmonic in M and

$$\frac{\partial g_1}{\partial N} = \frac{\partial g_2}{\partial N}$$

in ∂M, then $g_1 = g_2+$ const. in M.
d) If g is harmonic in M, then

$$\int_{\partial M} \frac{\partial g}{\partial N} \sigma = 0$$

e) The function $\frac{1}{(x^2+y^2+z^2)^{1/2}}$ is harmonic in $\mathbf{R}^3 - \{0\}$.
f) *(Mean value theorem)*. Let f be harmonic in the region

$$B_r = \{p \in \mathbf{R}^3; |p - p_0|^2 \leq r^2\}$$

whose boundary is the sphere S_r with center in p_0. Then

$$f(p_o) = \frac{1}{4\pi r^2} \int_{S_r} f\sigma$$

Hint: Use Green's second identity in the region $D = B_r - B_\rho$, $\rho < r$, with $f = f$ and $g = 1/r$. Since g and f are harmonic,

$$\int_{S_\rho} \left(f \frac{\partial}{\partial N} (\frac{1}{r}) - \frac{1}{r} \frac{\partial f}{\partial N} \right) \sigma = \int_{S_r} \left(f \frac{\partial}{\partial N} (\frac{1}{r}) - \frac{1}{r} \frac{\partial f}{\partial N} \right) \sigma.$$

Since $\frac{\partial}{\partial N}(1/r) = \frac{\partial}{\partial r}(1/r) = -1/r^2$, we obtain from (d),

$$\frac{1}{4\pi\rho^2} \int_{S_\rho} f\sigma = \frac{1}{4\pi r^2} \int_{S_r} f\sigma.$$

Now let $\rho \to 0$ to obtain the desired conclusion.
g) *(The maximum principle)*. Let f be a nonconstant harmonic function in a closed bounded region $M \subset \mathbf{R}^3$ (i.e., M is the union of a bounded connected open set with its boundary which is not necessarily regular). Then f reaches the maximum and the minimum in the boundary ∂M of M.

Hint: Assume that $f(p)$ is maximum, $p \in M - \partial M$ and consider a ball $B \subset M - \partial M$ with center p and such that $f(p) \geq f(q)$, for all $q \in B$. Show this contradicts (f).

12) Let M^n be a compact differentiable manifold without boundary. Show that M is orientable if and only if there exists a differential n-form ω defined on M and which is everywhere nonzero.

    Hint: For the "only if" part use a partition of unity to construct a nonzero n-form globally defined on M.

13) Let M be a compact, orientable, differentiable manifold without boundary. Show that M is not contractible to a point.

    Hint: Use Exercise 12, Poincaré's lemma and Stokes theorem.

14) Let A, B and C be differentiable functions in $\mathbf{R}^3$ and consider the differential system

$$\begin{cases} \frac{\partial R}{\partial y} - \frac{\partial Q}{\partial z} = A \\ \frac{\partial P}{\partial z} - \frac{\partial R}{\partial x} = B \\ \frac{\partial Q}{\partial x} - \frac{\partial P}{\partial y} = C \end{cases}$$

where P, Q and R are unknown functions in $\mathbf{R}^3$.

a) Show that a necessary and sufficient condition for a solution to the above system to exist is that

$$\frac{\partial A}{\partial x} + \frac{\partial B}{\partial y} + \frac{\partial C}{\partial z} = 0.$$

Hint: Consider in $\mathbf{R}^3$ the differential form

$$\omega = Ady \wedge dz + Bdz \wedge dx + Cdx \wedge dy$$

and notice that $d\omega = \left(\frac{\partial A}{\partial x} + \frac{\partial B}{\partial y} + \frac{\partial C}{\partial z}\right) dx \wedge dy \wedge dz$. By Poincaré's lemma, $d\omega = 0$ if and only if there exists a form $\alpha = Pdx + Qdy + Rdz$ with $d\alpha = \omega$; this last condition is precisely the above system.

b) Assume the above condition to be satisfied and determine the functions P, Q, R.

Hint: Consider the contraction $H(p, t) = tp$ of $\mathbf{R}^3$ to $(0, 0, 0)$. Then

$$\overline{\omega} = H^*\omega = A(tx, ty, tz)(ytdt \wedge dz - ztdt \wedge dy) + \ldots +$$
$$+ \text{ terms without } dt.$$

Thus the form α of (a) is given by

$$\alpha = I\overline{\omega} = \left(\int_0^1 A(tx, ty, tz)tdt\right)(ydz - zdy) + \ldots$$

15) Let v be a differentiable vector field in $\mathbf{R}^3$. Prove that:

a) If div $v = 0$, then there exists a vector field u in $\mathbf{R}^3$ such that rot $u = v$.

b) If rot $v = 0$, then there exists a function f in $\mathbf{R}^3$ such that grad $f = v$.

16) *(Brouwer fixed point theorem).*

a) Let M^n be a compact, orientable, differentiable manifold with boundary $\partial M \neq \phi$. Show that there exists no differentiable map $f : M \to \partial M$ such that the restriction $f|\partial M$ is the identity.

Hint (following E. Lima): Assume the existence of such an f, and let ω be the nonzero $(n-1)$-form on ∂M given by Exercise 12. Clearly $d(f^*\omega) = f^*(d\omega) = 0$, hence

$$0 = \int_M d(f^*\omega) = \int_{\partial M} i^* f^* \omega = \int_{\partial M} \omega \neq 0,$$

and this is a contradiction.

b) Prove the Brouwer fixed point theorem: Let $B \subset \mathbf{R}^n$ be the ball $\{p \in \mathbf{R}^n; |p| \leq 1\}$. Every differentiable map $g: B \to B$ has a fixed point, i.e., there exists $q \in B$ such that $g(q) = q$.

Hint: If $g(p) \neq p$, for all $p \in B$, the half-line starting in $g(p)$ and passing through p intersects ∂B in a unique point, say $q = f(p)$. The map $f: B \to \partial B$ so defined satisfy the conditions of (a) and yields a contradiction.

5. Differential Geometry of Surfaces

1. The Structure Equations of $\mathbf{R}^n$

We now apply our knowledge of differential forms to study some differential geometry. We start with a few definitions.

A *Riemannian manifold* is a differentiable manifold M and a choice, for each point $p \in M$, of a positive definite inner product $\langle\ ,\ \rangle_p$ in T_pM which varies differentiably with p in the following sense: If X and Y are differentiable vector fields in M, the function $p \mapsto \langle X, Y\rangle_p$ is differentiable in M. The inner product $\langle\ ,\ \rangle$ is usually called a *Riemannian metric* on M.

The notion of equivalence between Riemannian manifolds is the notion of *isometry*. A diffeomorphism $\varphi: M \to M'$ between Riemannian manifolds M and M' is an isometry if for all p and all pairs $x, y \in T_pM$, we have

$$\langle x, y\rangle_p = \langle d\varphi_p(x), d\varphi_p(y)\rangle_{\varphi(p)}.$$

The importance of the notion of Riemannian manifold is that we can define on it the usual metric notions (length, area, angles, etc.) of the euclidean geometry. Actually, euclidean geometry is just the study of metric notions in the simplest Riemannian geometry, namely, $\mathbf{R}^n$ endowed with the following inner product: If $x = (x_1, \ldots, x_n)$ and $y = (y_1, \ldots, y_n)$ are vectors in $\mathbf{R}^n$, one defines

$$\langle x, y\rangle = x_1y_1 + \ldots + x_ny_n.$$

Although $\mathbf{R}^n$ is the simplest Riemannian manifold, it is, in a certain sense, the universal Riemannian manifold. We hope to make this clearer later.

We will begin, therefore, by establishing the so-called structure equations of $\mathbf{R}^n$.

Let $U \subset \mathbf{R}^n$ be an open set and let $e_1, \ldots, e_n$ be n differentiable vector fields such that for each $p \in U$, $\langle e_i, e_j\rangle_p = \delta_{ij}$, where $\delta_{ij} = 0$ if $i \neq j$ and $\delta_{ij} = 1$ if $i = j$. Such a set of vector fields is called an *orthonormal moving frame*. From now on, we will omit the adjective orthonormal.

Given the moving frame $\{e_i\}$, $i = 1, \ldots, n$, we can define differential 1-forms ω_i by the condition $\omega_i(e_j) = \delta_{ij}$, $j = 1, \ldots, n$; in other words, at each p, the basis $\{(\omega_i)_p\}$ is the dual basis of $\{(e_i)_p\}$. The set of forms $\{\omega_i\}$ is called the *coframe associated to* $\{e_i\}$.

Each vector field e_i is a differentiable map $e_i: U \subset \mathbf{R}^n \to \mathbf{R}^n$. The differential at $p \in U$, $(de_i)_p: \mathbf{R}^n \to \mathbf{R}^n$, is a linear map. Thus, for each p and each $v \in \mathbf{R}^n$ we can write

$$(de_i)_p(v) = \sum_j (\omega_{ij})_p(v) e_j.$$

It is easily checked that the expressions $(\omega_{ij})_p(v)$, above defined, depend linearly on v. Thus $(\omega_{ij})_p$ is a linear form in $\mathbf{R}^n$ and, since e_i is a differentiable vector field, ω_{ij} is a differential 1-form. Keeping this in mind, we write the above as

$$de_i = \sum_j \omega_{ij} e_j. \tag{1}$$

The n^2 forms ω_{ij} so defined are called the *connection forms* of $\mathbf{R}^n$ in the moving frame $\{e_i\}$.

Not all of the forms ω_{ij} are independent. If we differentiate $\langle e_i, e_j \rangle = \delta_{ij}$, we obtain

$$0 = \langle de_i, e_j \rangle + \langle e_i, de_j \rangle = \omega_{ij} + \omega_{ji},$$

that is, the connection forms $\omega_{ij} = -\omega_{ji}$ are antisymmetric in the indices i, j.

The crucial point in the method of moving frames is that the forms ω_i and ω_{ij} satisfy the structure equations of Elie Cartan.

Proposition 1. *(The structure equations of* $\mathbf{R}^n$*). Let* $\{e_i\}$ *be a moving frame in an open set* $U \subset \mathbf{R}^n$*. Let* $\{\omega_i\}$ *be the coframe associated to* $\{e_i\}$ *and* ω_{ij} *the connection forms of* U *in the frame* $\{e_i\}$*. Then*

$$d\omega_i = \sum_k \omega_k \wedge \omega_{ki}, \tag{2}$$

$$d\omega_{ij} = \sum_k \omega_{ik} \wedge \omega_{kj}, \quad i, j, k = 1, \dots, n. \tag{3}$$

Proof. Let $a_1 = (1, \dots, 0), \dots, a_n = (0, \dots, 1)$ the canonical basis of $\mathbf{R}^n$, and let $x_i: U \to \mathbf{R}$ be the function that assigns to the point $(x_1, \dots, x_n)$ its ith-coordinate. Then dx_i is a differential 1-form on U and, since $dx_j(a_j) = \delta_{ij}$, we conclude that $\{dx_i\}$ is the coframe associated to $\{a_i\}$.

Now write

$$e_i = \sum_j \beta_{ij} a_j, \tag{4}$$

where β_{ij} is a differentiable function on U and, for each $p \in U$, the matrix $(\beta_{ij}(p))$ is an orthogonal matrix. Since $\omega_i(e_j) = \delta_{ij}$,

$$\omega_i = \sum_j \beta_{ij} dx_j. \tag{5}$$

We first prove that $d\beta_{ij} = \sum_k \omega_{ik}\beta_{kj}$. In fact,

$$de_i = \sum_k \omega_{ik} e_k = \sum_k \omega_{ik}(\sum_j \beta_{kj} a_j) = \sum_{jk} \omega_{ik}\beta_{kj} a_j,$$

and since from (4), $de_i = \sum d\beta_{ij} a_j$, we obtain by comparison,

$$d\beta_{ij} = \sum_k \omega_{ik}\beta_{kj}. \tag{6}$$

To obtain the first structure equation (2), we differentiate (5) and use (6):

$$d\omega_i = \sum_j d\beta_{ij} \wedge dx_j = \sum_{jk} \omega_{ik}\beta_{kj} \wedge dx_j = \sum_k \omega_k \wedge \omega_{ki}.$$

For the second structure equation (3), we differentiate (6), obtaining

$$0 = \sum_k d\omega_{ik}\beta_{kj} - \sum_k \omega_{ik} \wedge d\beta_{kj},$$

that is

$$\sum_k d\omega_{ik}\beta_{kj} = \sum_k \omega_{ik} \wedge \sum_s \omega_{ks}\beta_{sj},$$

or, finally, multiplying by the inverse matrix of (β_{kj})

$$d\omega_{ie} = \sum_k \omega_{ik} \wedge \omega_{ke},$$

as we wished. □

Remark 1. If we denote by $x\colon U \hookrightarrow \mathbf{R}^n$ the inclusion map, to say that the forms ω_i are dual to the frame $\{e_i\}$ is equivalent to saying that $dx = \sum \omega_i e_i$. Intuitively, the expressions that define ω_i and ω_{ij}, that is,

$$dx = \sum \omega_i e_i, \qquad de_i = \sum \omega_{ij} e_j,$$

describe how the moving frame $x, e_1, \ldots, e_n$ varies as we move (along a curve $x(t)$) in U. This was how Elie Cartan introduced the method of moving frames. The structure equations were then consequences of the "necessary" relations:

$$d(dx) = 0, \qquad d(de_i) = 0.$$

For instance, the first structure equation can be obtained as follows:

$$\begin{aligned} 0 = d(dx) &= \sum_i d\omega_i e_i - \sum \omega_i \wedge de_i \\ &= \sum_j d\omega_j e_j - \sum \omega_i \wedge \sum_j \omega_{ij} e_j = \sum_j (d\omega_j - \sum_i \omega_{ji} \wedge \omega_i) e_j, \end{aligned}$$

hence

$$d\omega_j = \sum_i \omega_i \wedge \omega_{ij}.$$

The second equation can be obtained similarly.

The main idea of Cartan's method to study the geometry of submanifolds of $\mathbf{R}^N$ can be described as follows. Let $x: M^n \to \mathbf{R}^{n+k}$ be an immersion of a differentiable manifold M^n into the euclidean space $\mathbf{R}^{n+k}$. It is a consequence of the inverse function theorem that for $p \in M$ there exists a neighborhood $U \subset M$ of p such that the restriction $x|U \subset M \to \mathbf{R}^{n+k}$ is an embedding. (See Exercise 4 of Chap. 3).

Let $V \subset \mathbf{R}^{n+k}$ be a neighborhood of $x(p)$ in $\mathbf{R}^{n+k}$ such that $V \cap M = x(U)$. Assume that V is such that there exists a moving frame $\{e_1, \ldots, e_n, e_{n+1}, \ldots, e_q\}$ in V with the property that, when restricted to $x(U)$, the vectors $e_1, \ldots, e_n$ are tangent to $x(U)$; such a moving frame is said to be an *adapted frame*.

In V we have, associated to the frame $\{e_i\}$, the coframe forms ω_i and the connection forms ω_{ij} which satisfy the structure equations (2) and (3). The map $x: U \subset M \to V \subset \mathbf{R}^{n+k}$ induce forms $x^*(\omega_i)$, $x^*(\omega_{ij})$ in U. Since x^* commutes with exterior derivation and exterior products, such forms in U satisfy again the structure equations (2) and (3). It turns out that the local metric geometry of $U \subset M$ is all contained in the structure equations, and this reflects the "universal character" of $\mathbf{R}^n$.

In the next section we will apply the method of moving frames to a simple but important case, namely, surfaces in $\mathbf{R}^3$. For that, we will need a few preliminary lemmas that we establish now.

Lemma 1. *(Cartan's lemma). Let V^n be a vector space of dimension n, and let $\omega_1, \ldots, \omega_r: V^n \to \mathbf{R}$, $r \leq n$, be linear forms in V that are linearly independent. Assume that there exist forms $\theta_1, \ldots, \theta_r: V \to \mathbf{R}$ such that $\sum_{i=1}^r \omega_i \wedge \theta_i = 0$. Then*

$$\theta_i = \sum_j a_{ij}\omega_j, \quad \text{with} \quad a_{ij} = a_{ji}.$$

Proof. We complete the forms ω_i into a basis $\omega_1, \ldots, \omega_r, \omega_{r+1}, \ldots, \omega_n$ of V^* and we write

$$\theta_i = \sum_j a_{ij}\omega_j + \sum_l b_{il}\omega_l, \quad l = r+1, \ldots, n.$$

By using the hypothesis, we obtain

$$0 = \sum_{i=1}^{n} \omega_i \wedge \theta_i = \sum_{ij} a_{ij}\omega_i \wedge \omega_j + \sum_{l} b_{il}\omega_i \wedge \omega_l$$
$$= \sum_{i<j}(a_{ij} - a_{ji})\omega_i \wedge \omega_j + \sum_{i<l} b_{il}\omega_i \wedge \omega_l.$$

Since $\omega_k \wedge \omega_s$, $k < s$, $k, s = 1, \ldots, n$, are linearly independent, we conclude that $b_{il} = 0$ and $a_{ij} = a_{ji}$. □

Lemma 2. *Let $U \subset \mathbf{R}^n$ and let $\omega_1, \ldots, \omega_n$ be linearly independent differential 1-forms in U. Assume that there exists a set of differential 1-forms $\{\omega_{ij}\}$, $i, j = 1, \ldots, n$ that satisfy the conditions:*

$$\omega_{ij} = -\omega_{ji}, \qquad d\omega_j = \sum \omega_k \wedge \omega_{kj}.$$

Then such a set is unique.

Proof. Suppose the existence of another set $\overline{\omega}_{ij}$ with

$$\overline{\omega}_{ij} = -\overline{\omega}_{ji}, \qquad d\omega_j = \sum_k \omega_k \wedge \overline{\omega}_{kj}.$$

Then

$$\sum_k \omega_k \wedge (\overline{\omega}_{kj} - \omega_{kj}) = 0,$$

and, by Cartan's lemma,

$$\overline{\omega}_{kj} - \omega_{kj} = \sum_i B^j_{ki}\omega_i, \quad B^j_{ki} = B^j_{ik}.$$

Notice that

$$\overline{\omega}_{kj} - \omega_{kj} = \sum_i B^j_{ki}\omega_i = -(\overline{\omega}_{jk} - \omega_{jk}) = -\sum B^k_{ji}\omega_i,$$

and, since the ω_i are linearly independent, $B^j_{ki} = -B^k_{ji}$. By using the above symmetries, we obtain finally that

$$B^k_{ji} = -B^j_{ki} = -B^j_{ik} = B^i_{jk} = B^i_{kj} = -B^k_{ij} = -B^k_{ji} = 0,$$

that is, $\overline{\omega}_{kj} = \omega_{kj}$. □

2. Surfaces in $\mathbf{R}^3$

We now apply the method of moving frames to the special case of surfaces in $\mathbf{R}^3$. Let $x: M^2 \to \mathbf{R}^3$ be an immersion of a two-dimensional differentiable manifold in $\mathbf{R}^3$. For each point $p \in M^2$, an inner product $\langle\ ,\ \rangle_p$ is defined in T_pM by the rule:

$$\langle v_1, v_2\rangle_p = \langle dx_p(v_1), dx_p(v_2)\rangle_{x(p)},$$

where the inner product in the right hand side is the canonical inner product of $\mathbf{R}^3$. It is straightforward to check that $\langle\ ,\ \rangle_p$ is differentiable and defines a Riemannian metric in M^2 to be called the *metric induced by the immersion* x.

We will study the local geometry of M^2 around a point $p \in M^2$. Let $U \subset M$ be a neighborhood of p such that the restriction $x|U$ is an embedding. Let $V \subset \mathbf{R}^3$ be a neighborhood of p is $\mathbf{R}^3$ such that $V \cap x(M) = x(U)$, and that it is possible to choose in V an adapted moving frame e_1, e_2, e_3; this means that, when restricted to $x(U)$, e_1 and e_2 are tangent to $x(U)$ (hence e_3 is normal to $x(U)$).

In V we have, associated to the frame $\{e_i\}$, the coframe forms ω_i and the connection forms $\omega_{ij} = -\omega_{ji}$, $i,j = 1,2,3$, which satisfy the structure equations:

$$\begin{aligned}
d\omega_1 &= \omega_2 \wedge \omega_{21} + \omega_3 \wedge \omega_{31},\\
d\omega_2 &= \omega_1 \wedge \omega_{12} + \omega_3 \wedge \omega_{32},\\
d\omega_3 &= \omega_1 \wedge \omega_{13} + \omega_2 \wedge \omega_{23},\\
d\omega_{12} &= \omega_{13} \wedge \omega_{32},\\
d\omega_{13} &= \omega_{12} \wedge \omega_{23},\\
d\omega_{23} &= \omega_{21} \wedge \omega_{13}.
\end{aligned}$$

The immersion $x: U \subset M \to V \subset \mathbf{R}^3$ induces forms $x^*(\omega_i)$, $x^*(\omega_{ij})$ in U. Since x^* commutes with d and $\wedge$, such forms still satisfy the structure equations. Notice that $x^*(\omega_3) = 0$, since for all $q \in U$ and all $v \in T_qM$,

$$x^*(\omega_3)(v) = \omega_3(dx(v)) = \omega_3(a_1e_1 + a_2e_2) = 0,$$

where $v = a_1e_1 + a_2e_2$.

With a slight abuse of notation, we will write

$$x^*(\omega_i) = \omega_i, \qquad x^*(\omega_{ij}) = \omega_{ij}.$$

This amounts to look upon U as a subset of $\mathbf{R}^3$ by the inclusion $x: U \to \mathbf{R}^3$ (notice that $x|U$ is an embedding), and to look upon the forms ω_i and ω_{ij} as restricted to U. These restricted forms satisfy the above structure equations with the additional relation $\omega_3 = 0$.

Since $\omega_3 = 0$,

$$d\omega_3 = \omega_1 \wedge \omega_{13} + \omega_2 \wedge \omega_{23} = 0,$$

hence, by Cartan's lemma,

$$\omega_{13} = h_{11}\omega_1 + h_{12}\omega_2,$$
$$\omega_{23} = h_{21}\omega_1 + h_{22}\omega_2,$$

where $h_{ij} = h_{ji}$ are differentiable functions in U.

We want to obtain a geometric interpretation of the functions h_{ij}. For that, observe that the map $e_3 : U \to \mathbf{R}^3$ takes its values in the unit sphere $S^2 \subset \mathbf{R}^3$, since $|e_3| = 1$. By fixing orientations of U and $\mathbf{R}^3$, we can choose the frame $\{e_i\}$ in such a way that, for each $q \in U$, $\{e_1, e_2\}$ is in the orientation of U and $\{e_1, e_2, e_3\}$ is in the orientation of $\mathbf{R}^3$. In this case, $e_3 : U \to S^2 \subset \mathbf{R}^3$ is well defined, does not depend on the choice of the frame, and it is called the *Gauss* (normal) *map* in U (Fig. 5.1).

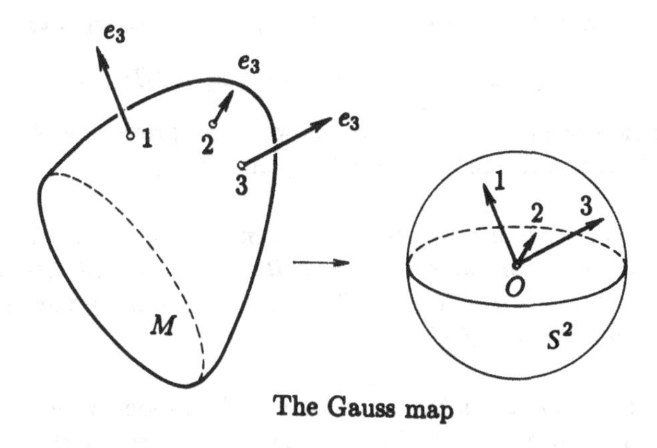

The Gauss map

Fig. 5.1

We notice the important fact that if M is oriented, the Gauss map can be defined globally on M.

Now, since $de_3 = \omega_{31}e_1 + \omega_{32}e_2$, we obtain, for all $q \in U$ and all $v = a_1e_1 + a_2e_2 \in T_Qm$,

$$de_3(v) = -\begin{pmatrix} h_{11} & h_{12} \\ h_{21} & h_{22} \end{pmatrix} \begin{pmatrix} a_1 \\ a_2 \end{pmatrix},$$

that is, $(-h_{ij})$ is the matrix of the differential of the Gauss map $e_3 : U \to S^2$ in the basis $\{e_1, e_2\}$ which is the geometric interpretation we were looking for.

Since the matrix (h_{ij}) is symmetric, we conclude immediately that the differential $de_3 : TM \to TS^2$ of the Gauss map $e_3 : U \to S^2$ is a selfadjoint linear map. From a well known result in Linear Algebra, we know that such a linear map can be diagonalized with real eigenvalues $-\lambda_1, -\lambda_2$ and orthogonal eigenvectors.

It is usual to define the *Gaussian curvature* K of M in p by

$$K = \det(de_3)_p = \lambda_1\lambda_2 = h_{11}h_{22} - h_{12}^2$$

and the *mean curvature* H of M at p by

$$H = -\frac{1}{2}(\text{trace } de_3)_p = \frac{\lambda_1 + \lambda_2}{2} = \frac{h_{11} + h_{22}}{2},$$

where the functions involved are computed at p. Clearly K and H do not depend on the choice of the moving frame. Notice that H changes sign with a change of orientation but K remains the same under such a change. The expressions of K and H in terms of a moving frame are immediately obtained:

$$d\omega_{12} = \omega_{13} \wedge \omega_{32} = -(h_{11}h_{22} - h_{12}^2)\omega_1 \wedge \omega_2 = -K\omega_1 \wedge \omega_2,$$

$$\omega_{13} \wedge \omega_2 + \omega_1 \wedge \omega_{23} = (h_{11} + h_{22})\omega_1 \wedge \omega_2 = 2H\omega_1 \wedge \omega_2.$$

The expression $d\omega_{12} = -K\omega_1 \wedge \omega_2$ allows us to prove one of the most important theorems in the theory of surfaces in $\mathbf{R}^3$.

Theorem 1. *(Gauss). K only depends on the induced metric of M^2; that is, if $x, x': M^2 \to \mathbf{R}^3$ are two immersions with the same induced metrics, then $K(p) = K'(p)$, $p \in M$, where K and K' are the Gaussian curvatures of the immersions x and x', respectively.*

Proof. Let $U \subset M$ be a neighborhood of p and consider a moving frame $\{e_1, e_2\}$ in U, orthonormal in the induced metric. The set $\{dx(e_1), dx(e_2)\}$ can be extended into an adapted frame in $V \supset x(U)$ and, similarly, the set $\{dx'(e_1), dx'(e_2)\}$ can be extended into an adapted frame in $V' \supset x'(U)$.

Let us denote by a prime the entities that refer to the immersion x'. Then, $\omega_1 = \omega_1'$, $\omega_2 = \omega_2'$, by duality. By the uniqueness of Lemma 2, $\omega_{12} = \omega_{12}'$. It follows that

$$d\omega_{12} = d\omega_{12}' = -K\omega_1 \wedge \omega_2 = -K'\omega_1 \wedge \omega_2,$$

hence $K = K'$. □

Gauss theorem means that the Gaussian curvature, the definition of which made use of the ambient space $\mathbf{R}^3$, only depends on measurements made on the surface. This led Gauss, around 1827, to imagine the existence of geometries that were independent of the ambient space. Because he lacked adequate tools (in particular, the notion of a differentiable manifold), Gauss did not develop these ideas which were later (1852) taken up by Riemann. In general, geometric entities on M that can be computed from ω_1, ω_2 and ω_{12} depend only on the induced metric in the sense above described, and we ought to be able to define them with no mention to the immersion x. We will come back to that in the next section.

Example 1. Consider the immersion $x: U \subset \mathbf{R}^2 \to \mathbf{R}^3$, where U is

$$U = \{(s, v) \in \mathbf{R}^2; -\infty < s < \infty, 0 < v < 2\pi\},$$

and x is given by

$$x(x, v) = (h(s)\sin v, h(s)\cos v, g(s)).$$

Here $h(s) \neq 0$ and $g(s)$ are differentiable functions that satisfy

$$\left(\frac{dh}{ds}\right)^2 + \left(\frac{dg}{ds}\right)^2 = 1.$$

The image $x(U)$ is a *rotation surface* with axis Oz whose generating curve $y = h(s)$, $z = g(s)$ is parametrized by the arc length s (Fig. 5.2).

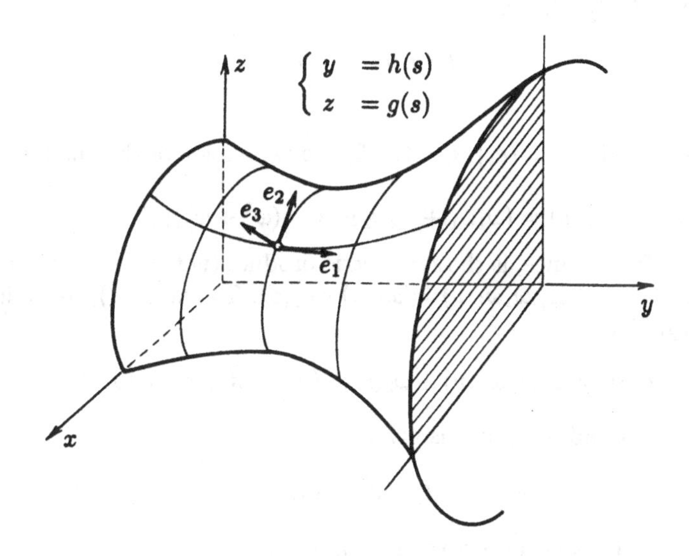

Fig. 5.2

We want to show that the Gaussian curvature of this surface is $K = -(h''/h)$, where prime denotes derivative relative to s.

Observe that v/h measures the length of the parallel $x(\text{const.}, v)$. Thus,

$$e_1 = dx\left(\frac{\partial}{\partial s}\right), \qquad e_2 = dx\left(\frac{1}{h}\frac{\partial}{\partial v}\right)$$

are orthonormal vectors tangent to $x(U)$. Together with a unit vector e_3 normal to $x(U)$, they constitute a frame adapted to the immersion x.

It is immediate to check that $\omega_1 = ds$, $\omega_2 = hdv$. On the other hand, if we set $\omega_{12} = ads + bdv$, we obtain

$$bds \wedge dv = ds \wedge \omega_{12} = \omega_1 \wedge \omega_{12} = d\omega_2 = dh \wedge dv = h'ds \wedge dv$$

and

$$ahds \wedge dv = \omega_{12} \wedge hdv = \omega_{12} \wedge \omega_2 = d\omega_1 = 0.$$

It follows that $\omega_{12} = h'dv$, and

$$d\omega_{12} = h''ds \wedge dv = \frac{h''}{h}\omega_1 \wedge \omega_2 = -K\omega_1 \wedge \omega_2,$$

and this yields the required expression. In particular, for the sphere of radius R, we have $h(s) = R\cos(s/R)$, hence $K = 1/R^2$.

As we mentioned before, to a given immersion $x\colon M^2 \to \mathbf{R}^2$ we associate two quadratic forms at each T_pM, $p \in M$, which are defined as follows.

The *first quadratic form* I_p is merely the quadratic form associated to the bilinear form $\langle\ ,\ \rangle_p$, that is,

$$I_p(v, v) = \langle v, v\rangle_p$$

In an adapted frame $\{e_i\}$, $i = 1, 2, 3$, the expression of the quadratic form I is given by

$$I(v) = (\omega_1\omega_1 + \omega_2\omega_2)(v) = (\omega_1^2 + \omega_2^2)(v), \tag{7}$$

where $\omega_1\omega_1$, for instance, is the symmetric product (not the exterior product) of ω_1 with ω_1, that is, $\omega_1\omega_1(v) = \omega_1(v) \cdot \omega_1(v)$. To check (7), we write $v = a_1e_1 + a_2e_2$. Then

$$I_p(v) = \omega_1(v)\omega_1(v) + \omega_2(v)\omega_2(v) = a_1^2 + a_2^2 = \langle v, v\rangle.$$

Thus, the first quadratic form is given by

$$I = \omega_1^2 + \omega_2^2.$$

where, as usually, we dropped the indication of p.

The *second quadratic form* is defined in an adapted moving frame by

$$II_p(v) = (\omega_{13}\omega_1 + \omega_{23}\omega_2)(v) = \sum_{ij} h_{ij}\omega_i\omega_j, \quad i, j = 1, 2,$$

where again we are considering symmetric products of differential forms. In this case, we have to prove that II does not depend on the choice of frames. This is actually so, since this is the quadratic form associated to minus the differential of the Gauss map, that is,

$$II_p(v) = -\langle de_3(v), v\rangle_p, \quad v \in T_pM.$$

It is convenient to describe still another interpretation of II that can be obtained as follows. Let $\alpha: (-\varepsilon, \varepsilon) \to M$ be a curve in M parametrized by the arclength s, with $\alpha(0) = p$, $\alpha'(0) = v \in T_pM$. Then, by writing $x \circ \alpha(s) = x(s)$ and $e_3 \circ \alpha(s) = e_3(s)$, we obtain

$$\langle \frac{dx}{ds}, e_3(s)\rangle = 0,$$

hence

$$\begin{aligned}\langle \frac{d^2x}{ds^2}, e_3(s)\rangle\Big|_{s=0} &= -\langle \frac{dx}{ds}, \frac{de_3}{ds}\rangle\Big|_{s=0} = -\langle dx(v), de_3(v)\rangle_p \\ &= \langle \omega_1 e_1 + \omega_2 e_2, \omega_{31} e_1 + \omega_{32} e_2\rangle(v) \\ &= (\omega_1\omega_{13} + \omega_2\omega_{23})(v) = II_p(v).\end{aligned}$$

On the other hand, denoting by $k(s)$ the curvature of the curve $\alpha(s)$ and by $n(s)$ the principal normal of $\alpha(s)$, we obtain that

$$\langle \frac{d^2x}{ds}(0), e_3(0)\rangle = k(0)\langle n(0), e_3(0)\rangle.$$

The expression $k\langle n, e_3\rangle(p)$ is called the *normal curvature* $k_n(v)$ of the surface in the direction $v = \alpha'(0)$ at the point p. Since $II_p(v) = k_n(v)$, we have that $k_n(v)$ is the same for all curves $\alpha(s)$ with the same tangent vector v at p.

Thus, collecting the two interpretations, we conclude that

$$II_p(v) = -\langle de_3(v), v\rangle_p = k_n(v).$$

It is known from Linear Algebra that the maximum and minimum of $II_p(v)$, as v runs the unit circle $S^1 \subset T_pS$, are the eigenvalues $-\lambda_1, -\lambda_2$ of $(-de_3)$ and the corresponding vectors generate the eigenspaces of $(-de_3)$. The extremal normal curvatures $(-\lambda_1) = k_1$, $(-\lambda_2) = k_2$ are called the *principal curvatures* at p and the corresponding directions are called the *principal directions* at p.

The importance of the first and second fundamental forms is that they determine the local geometry of surfaces in $\mathbf{R}^3$.

In the same vein, we can work out the whole local geometry of surfaces in $\mathbf{R}^3$ (See, for instance, [dC] Chapter 3). We stop here, however, in the hope that the method is sufficiently clear. It only remains to be explained what is meant by the statement that I and II determine entirely the local geometry of a surface in $\mathbf{R}^3$. This is the content of the next theorem, and the corollary following it.

Theorem 2. *Let U and U' be connected submanifolds of dimension two in $\mathbf{R}^3$. Assume that there exist adapted frames $\{e_i\}$ in U, $\{e_i'\}$ in U', $i = 1, 2, 3$, and a diffeomorphism $f: U \to U'$ such that*

$$f^*\omega_i' = \omega_i, \quad f^*\omega_{ij}' = \omega_{ij} \quad j = 1, 2, 3.$$

Then there exists a rigid motion $\rho: \mathbf{R}^3 \to \mathbf{R}^3$ such that the restriction $\rho|U = f$.

Proof. Let $p \in U$ and $f(p) \in U'$. Let T be the translation of $\mathbf{R}^3$ that takes p to $p' = f(p)$ and let R be the rotation of $\mathbf{R}^3$ that takes e_i to e_i'. Set $\rho = R \circ T$. We will show that $g = f \circ \rho^{-1}: \rho(U) \to U'$ is the identity in U', and this implies the statement of the theorem.

Since ρ is an isometry of $\mathbf{R}^3$, consider the orthonormal moving frame $\tilde{e}_i = d\rho(e_i)$ in $\rho(U)$. We will denote with an upper index $\sim$ the entities associated to the frame $\{\tilde{e}_i\}$ in $\rho(U)$. By definition, for all $q \in \rho(U)$ and all $v \in T_q(\rho(U))$,

$$(d\tilde{e}_i)_q(v) = \sum_j (\tilde{\omega}_{ij})_q(v)(\tilde{e}_j)_q.$$

Define $e_i' \circ g$ by $(e_i' \circ g)(q) = e_i'(g(q))$. Then

$$\begin{aligned} d(e_i' \circ g)_q(v) &= (de_i')_{g(q)}(dg(v)) = \sum_j (\omega_{ij}')_{g(q)}(dg(v))(e_j')_{g(q)} \\ &= \sum_j (g^*\omega_{ij}')_q(v)(e_j' \circ g)_q = \sum_j (\tilde{\omega}_{ij})_q(v)(e_j' \circ g)_q, \end{aligned}$$

where the last equality follows from the fact that

$$g^*\omega_{ij}' = (f \circ \rho^{-1})^*\omega_{ij}' = (\rho^{-1})^* f^*\omega_{ij}' = (\rho^{-1})^*\omega_{ij} = \tilde{\omega}_{ij}.$$

Because q and v are arbitrary, it follows that $\tilde{e}_i - e_i' \circ g$ satisfies the system of ordinary differential equations

$$d(\tilde{e}_i - e_i' \circ g) = \sum_j \tilde{\omega}_{ij}(\tilde{e}_j - e_j' \circ g),$$

with initial conditions at the point $\rho(p)$ given by:

$$(\tilde{e}_i - e_i' \circ g)(\rho(p)) = 0.$$

By the uniqueness theorem for ordinary differential equations, $\tilde{e}_i = e_i' \circ \rho$.

In a similar way, we can show that

$$d(\tilde{x} - x' \circ g) = \sum_i \tilde{\omega}_i(\tilde{e}_i - e_i' \circ g) = 0,$$

where $\tilde{x}: \rho(U) \subset \mathbf{R}^3$ and $x': U' \subset \mathbf{R}^3$ are the respective inclusions and the last equality comes from what we just proved. Since the initial conditions in $\rho(p)$ are: $(\tilde{x} - x' \circ g)(\rho(p)) = 0$, we conclude that $\tilde{x} = x' \circ g$. Since $\tilde{x}$ and x' are inclusions, this implies that g is the identity, as we wished. □

Corollary. *Let U and U' be connected submanifolds of dimension two in $\mathbf{R}^3$. Assume that there exists a diffeomorphism $f: U \to U'$ which preserves the first and second quadratic forms, that is,*

$$I_p(v,v) = I'_{f(p)}(df(v), df(v)), \quad II_p(v,v) = II'_{f(p)}(df(v), df(v))$$

for all $p \in U$ and all $v \in T_pU$. Then, there exists a rigid motion $\rho: \mathbf{R}^3 \to \mathbf{R}^3$ such that $\rho|U = f$.

Proof. Consider in U an adapted frame $\{e_i\}$ and define in U' a frame $\{e'_i\} = \{df(e_i)\}$. Since f preserves inner products, this is again an adapted frame, and $f^*\omega_i = \omega_i$. Because the second fundamental forms are preserved, $(h_{ij}) = (h'_{ij} \circ f)$. Thus, $f^*\omega'_{13} = \omega_{13}$ and $f^*\omega'_{23} = \omega_{23}$. Finally, by Lemma 2 (uniqueness of connection forms), we see that $f^*\omega_{12} = \omega_{12}$. We can now apply the theorem to obtain the conclusion.

3. Intrinsic Geometry of Surfaces

In the study of surfaces M^2 in $\mathbf{R}^3$ we have seen that certain geometric entities, for instance, the Gaussian curvature, only depend on the first fundamental form, that is to say, on the Riemannian metric of M^2. A surprising number of geometric properties of surfaces are in the same situation as the Gaussian curvature, i.e., they only depend on the first fundamental form, and they constitute the intrinsic geometry of surfaces. In this section, we will present a more systematic study of such properties by using the method of moving frames.

Our starting point is a two-dimensional differentiable manifold M^2 together with a Riemannian metric $\langle\ ,\ \rangle$. For each point $p \in M$, choose a neighborhood $U \subset M$ of p such that one can define orthonormal vector fields e_1 and e_2 on U. From this moving frame $\{e_1, e_2\}$, we can define a corresponding coframe $\{\omega_1, \omega_2\}$ by the condition $\omega_i(e_j) = \delta_{ij}$, $i, j = 1, 2$. The question now is whether we can define differential forms that play the role of connection forms.

The choice we make below can be motivated by the following considerations. If U could be isometrically embedded (that is, in such a way that the Riemannian inner product $\langle\ ,\ \rangle$ in M^2 is induced by $\mathbf{R}^3$), we would obtain a moving frame e_1, e_2, e_3 in a open set $V \supset U$ of $\mathbf{R}^3$ that extends the frame e_1, e_2 in U. From the forms $\omega_1, \omega_2, \omega_{12}, \omega_{13}, \omega_{23}$, and from the structure equations

$$\begin{aligned} d\omega_1 &= \omega_{12} \wedge \omega_2, \\ d\omega_2 &= \omega_{21} \wedge \omega_1, \\ d\omega_{12} &= \omega_{13} \wedge \omega_{32}, \\ d\omega_{13} &= \omega_{12} \wedge \omega_{23}, \\ d\omega_{23} &= \omega_{21} \wedge \omega_{13}, \end{aligned}$$

only the forms $\omega_1, \omega_2, \omega_{12}$ and the first two equations do not contain elements related to the "external" vector e_3. It is thus reasonable to expect that there exists in U a unique form $\omega_{12} = -\omega_{21}$ such that the two first equations hold. This is indeed the case.

Lemma 3. *(Theorem of Levi-Civitta). Let M^2 be a Riemannian (two-dimensional) manifold. Let $U \subset M$ be an open set where a moving orthonormal frame $\{e_1, e_2\}$ is defined, and let $\{\omega_1, \omega_2\}$ be the associated coframe. Then there exists a unique 1-form $\omega_{12} = -\omega_{21}$ such that*

$$d\omega_1 = \omega_{12} \wedge \omega_2, \quad d\omega_2 = \omega_{21} \wedge \omega_1.$$

Proof. Uniqueness has already been proved in Lemma 2 of Section 1. To prove existence, just define

$$\omega_{12}(e_1) = d\omega_1(e_1, e_2),$$
$$\omega_{12}(e_2) = d\omega_2(e_1, e_2),$$

and check the required properties: For instance,

$$d\omega_1(e_1, e_2) = \omega_{12}(e_1) = \omega_{12}(e_1)\omega_2(e_2) - \omega_{12}(e_2)\omega_2(e_1) = (\omega_{12} \wedge \omega_2)(e_1, e_2). \square$$

The problem now is to obtain geometric entities (that is, independent of the choice of frame) from the forms $\omega_1, \omega_2, \omega_{12}$. For that, it is convenient to see how such forms change under a change of frame.

Let $\{\bar{e}_1, \bar{e}_2\}$ be another frame in U. If $\{\bar{e}_1, \bar{e}_2\}$ has the same orientation as $\{e_1, e_2\}$, we obtain

$$\bar{e}_1 = f e_1 + g e_2$$
$$\bar{e}_2 = -g e_1 + f e_2,$$

where f and g and differentiable functions in U, and $f^2 + g^2 = 1$; on the other hand, if the orientations of $\{\bar{e}_1, \bar{e}_2\}$ and $\{e_1, e_2\}$ are opposite, we obtain

$$\bar{e}_1 = f e_1 + g e_2$$
$$\bar{e}_2 = g e_1 - f e_2.$$

Lemma 4. *If $\{\bar{e}_1, \bar{e}_2\}$ and $\{e_1, e_2\}$ have the same orientation, then*

$$\omega_{12} = \bar{\omega}_{12} - \tau,$$

where $\tau = f dg - g df$. If the above orientations are opposite,

$$\omega_{12} = -\bar{\omega}_{12} - \tau.$$

Proof. If the orientations are the same, we obtain that

$$\omega_1 = f\bar{\omega}_1 - g\bar{\omega}_2, \tag{1}$$

$$\omega_2 = g\overline{\omega}_1 + f\overline{\omega}_2. \tag{2}$$

Differentiating (1), we obtain

$$d\omega_1 = df \wedge \overline{\omega}_1 + f d\overline{\omega}_1 - dg \wedge \overline{\omega}_2 - g d\overline{\omega}_2.$$

By using the structure equations for $d\overline{\omega}_1$ and $d\overline{\omega}_2$, and the fact that $\overline{\omega}_{12} = -\overline{\omega}_{21}$, it follows that

$$d\omega_1 = \overline{\omega}_{12} \wedge \omega_2 + (f df + g dg) \wedge \omega_1 + (g df - f dg) \wedge \omega_2.$$

Since $f^2 + g^2 = 1$, $f df + g dg = 0$. Thus,

$$d\omega_1 = \overline{\omega}_{12} \wedge \omega_2 - \tau \wedge \omega_2 = (\overline{\omega}_{12} - \tau) \wedge \omega_2.$$

Similarly, by differentiating (2), we obtain

$$d\omega_2 = -(\overline{\omega}_{12} - \tau) \wedge \omega_1.$$

By the uniqueness of the connection form, we conclude finally that

$$\omega_{12} = \overline{\omega}_{12} - \tau,$$

and this proves the first part of the lemma. The case in which the orientations are opposite is analogous. □

A geometric interpretation for the 1-form τ is given below and asserts that, along a curve in U, τ is the differential of the "angle function" between e_1 and $\overline{e}_1$ along the curve; actually, what we will do is to show that it is possible to define such a function in a way that it is differentiable.

Lemma 5. *Let $p \in U \subset M$ be a point and let $\gamma: I \to U$ be a curve such that $\gamma(t_0) = p$. Let $\varphi_0 =$ angle $(e_1(p), \overline{e}_1(p))$. Then*

$$\varphi(t) = \int_{t_0}^{t} \left(f\frac{dg}{dt} - g\frac{df}{dt} \right) dt + \varphi_0$$

is a differentiable function such that:

$$\cos\varphi(t) = f, \quad \operatorname{sen}\varphi(t) = g, \quad \varphi(t_0) = \varphi_0, \quad d\varphi = \gamma^*\tau.$$

Proof. We first show that

$$f(t)\cos\varphi(t) + g(t)\operatorname{sen}\varphi(t) \equiv 1. \tag{3}$$

To see that, notice that from the definition of φ, we have that $\varphi' = fg' - gf'$. Thus,

$$\begin{aligned}(f\cos\varphi + g\operatorname{sen}\varphi)' &= f'\cos\varphi - f\operatorname{sen}\varphi\varphi' + g'\operatorname{sen}\varphi + g\cos\varphi\varphi' \\ &= (g' + fgf' - f^2g')\operatorname{sen}\varphi + (f' - g^2f' + gfg')\cos\varphi = 0,\end{aligned}$$

where in the last equality we have used that, since $f^2 + g^2 = 1$, $ff' + gg' = 0$. Therefore, $f \cos\varphi + g \operatorname{sen}\varphi = \text{const.}$, and since

$$f(t_0)\cos\varphi(t_0) + g(t_0)\operatorname{sen}\varphi(t_0) = (f^2 + g^2)(t_0) = 1$$

we conclude (3).

It follows that

$$(f - \cos\varphi)^2 + (g - \operatorname{sen}\varphi)^2 = f^2 + g^2 - 2f\cos\varphi - 2g\operatorname{sen}\varphi + 1 = 0,$$

hence

$$\cos\varphi(t) = f(t), \operatorname{sen}\varphi(t) = g(t),$$

and the lemma follows immediately. □

We are now in a position to develop the intrinsic geometry of surfaces. The first observation is that from (1) and (2) it follows that in a oriented surface the 2-form

$$\omega_1 \wedge \omega_2 = \overline{\omega}_1 \wedge \overline{\omega}_2 = \sigma$$

does not depend on the choice of frames and is therefore globally defined in M^2. The geometric meaning of the form σ is obtained as follows. If $v_1 = a_{11}e_1 + a_{12}e_2$, $v_2 = a_{21}e_1 + a_{22}e_2$ are linearly independent vectors at a point $p \in M$, then

$$\sigma(v_1, v_2) = \det(a_{ij}) = \text{area } (v_1, v_2),$$

where (v_1, v_2) denotes the parallelogram generated by v_1 and v_2. Because of that, σ is called the *area element* of M.

The next object of intrinsic geometry is motivated by Gauss theorem.

Proposition 2. *Let M^2 be a Riemannian manifold of dimension two. For each $p \in M$, we define a number $K(p)$ by choosing a moving frame $\{e_1, e_2\}$ around p and setting*

$$d\omega_{12}(p) = -K(p)(\omega_1 \wedge \omega_2)(p).$$

Then $K(p)$ does not depend on the choice of frames, and it is called the Gaussian curvature of M at p.

Proof. Let $\{\bar{e}_1, \bar{e}_2\}$ be another moving frame around p. Assume first that the orientations of the two moving frames are the same. Then

$$\omega_{12} = \overline{\omega}_{12} - \tau.$$

Since $\tau = fdg - gdf$, $d\tau = 0$, hence $d\omega_{12} = d\overline{\omega}_{12}$. It follows that

$$-K\omega_1 \wedge \omega_2 = d\omega_{12} = d\overline{\omega}_{12} = -\overline{K}\overline{\omega}_1 \wedge \overline{\omega}_2 = -\overline{K}\omega_1 \wedge \omega_2$$

hence $K = \overline{K}$, as we wished.

If the orientations an opposite, we obtain

$$d\omega_{12} = -d\overline{\omega}_{12}, \quad \omega_1 \wedge \omega_2 = -\overline{\omega}_1 \wedge \overline{\omega}_2$$

and the same conclusion holds. □

Another entity that does not depend on the choice of frames is the (covariant) derivative of vectors.

Definition 1. Let M^2 be a Riemannian manifold and let Y be a differentiable vector field on M. Let $p \in M$, $x \in T_pM$, and consider a curve $\alpha: (-\varepsilon, \varepsilon) \to M$ with $\alpha(0) = p$, $\alpha'(0) = x$. To define the *covariant derivative* $(\nabla_x Y)(p)$ of Y *relative to* x *in* p, we choose a moving frame $\{e_i\}$ around p, express $Y(\alpha(t))$ in this frame

$$Y(\alpha(t)) = \sum y_i(t) e_i, \quad i = 1, 2,$$

and set

$$(\nabla_x Y)(p) = \sum_i \left(\frac{dy_i}{dt}(0) + \sum_j \omega_{ji}(x) y_j(0) \right) e_i, \quad i, j = 1, 2,$$

where the convention is made that $\omega_{ii} = 0$.

Lemma 6. *The covariant derivative does not depend on the choice of frames.*

Proof. Let $\{e_1, e_2\}$ and $\{\overline{e}_1, \overline{e}_2\}$ be two orthonormal frames around p. Assume that they have the same orientation. Then

$$\begin{cases} y_1 & = f\overline{y}_1 - g\overline{y}_2 \\ y_2 & = g\overline{y}_1 + f\overline{y}_2 \end{cases} \qquad \begin{cases} e_1 & = f\overline{e}_1 - g\overline{e}_2 \\ e_2 & = g\overline{e}_1 + f\overline{e}_2 \end{cases} \tag{5}$$

where $Y(\alpha(t)) = \sum y_i(t) e_i = \sum \overline{y}_i(t) \overline{e}_i$, and f, g are differentiable functions with $f^2 + g^2 = 1$. By definition,

$$\nabla_x Y = \left(\frac{dy_1}{dt} + \omega_{21}(x) y_2 \right) e_1 + \left(\frac{dy_2}{dt} + \omega_{12}(x) y_1 \right) e_2$$

where the functions are taken at $t = 0$. By using (5), and the facts that $\omega_{12} = \overline{\omega}_{12} - \tau$ and $ff' + gg' = 0$, we arrive after a long but straightforward computation that

$$\nabla_x Y = \left(\frac{d\overline{y}_1}{dt} + \overline{\omega}_{21}(x) \overline{y}_2 \right) \overline{e}_1 + \left(\frac{d\overline{y}_2}{dt} + \overline{\omega}_{12}(x) \overline{y}_1 \right) \overline{e}_2$$

which proves the lemma in this case.

When the orientations of the frames are opposite, the proof is similar. □

The notion of covariant derivative can be used to give a geometric interpretation of the connection form ω_{12} associated to a moving frame $\{e_1, e_2\}$. In fact, since $e_1 = 1 \cdot e_1 + 0 e_2$, we obtain $\nabla_x e_1 = \omega_{12}(x) e_2$, hence

$$\omega_{12}(x) = \langle \nabla_x e_1, e_2 \rangle.$$

Thus the form ω_{12} applied to a vector x is the e_2-component of the covariant derivative $\nabla_x e_1$.

Remark. The covariant derivative was introduced by Levi-Civitta in 1916. For the induced metric of surfaces $M^2 \subset \mathbf{R}^3$, it can be shown (See Exercise 7) that the covariant derivative $\nabla_x Y$ is just the projection onto the tangent plane of M of the usual derivative in $\mathbf{R}^3$ of Y along a curve tangent to x. Thus, on a certain sense, $\nabla_x Y$ is the derivative of Y as "seen from the surface".

Starting from the covariant derivative, we can develop all concepts of the Riemannian geometry in dimension two (parallelism, geodesics, geodesic curvature, etc; see, for instance, [dC] Chapter 4). In what follows, we will present a short exposition of these ideas. M^2 will always be a two-dimensional Riemannian manifold.

Definition 2. A vector field Y along a curve $\alpha: I \to M^2$ is said to be *parallel along* α if $\nabla_{\alpha'(t)} Y = 0$, for all $t \in I$.

Definition 3. A curve $\alpha: I \to M^2$ is a *geodesic* if $\alpha'(t)$ is a parallel field along α.

Definition 4. Assume that M^2 is oriented, and let $\alpha: I \to M$ be a differentiable curve parametrized by the arc length s with $\alpha'(s) \neq 0$, $s \in I$. In a neighborhood of a point $\alpha(s) \in M$, consider a moving frame $\{e_1, e_2\}$ in the orientation of M such that, restricted to α, $e_1(s) = \alpha'(s)$. The *geodesic curvature* k_g of α in M is defined by

$$k_g = (\alpha^* \omega_{12})(\frac{d}{ds}),$$

where $\frac{d}{ds}$ is the canonical basis of $\mathbf{R}$.

Proposition 3. *Let $\alpha: I \to M^2$ and $\{e_1, e_2\}$ be as in Definition 4 (here we don't need to assume that M^2 is orientable, so that there are two possible choices for e_2). Then e_1 is parallel along α if and only if $\alpha^* \omega_{12} = 0$.*

Proof. e_1 is parallel along α if and only if $\nabla_{e_1} e_1 = 0$. Since $\langle \nabla_{e_1} e_1, e_1 \rangle = 0$, the last statement is equivalent to

$$0 = \langle \nabla_{e_1} e_1, e_2 \rangle = \omega_{12}(e_1),$$

or to $\alpha^* \omega_{12} = 0$, as we wished. □

Corollary. *A differentiable curve $\alpha: I \to M^2$ is a geodesic if and only if its geodesic curvature vanishes everywhere.*

A geometric interpretation of the geodesic curvature is given below.

Proposition 4. *Let M^2 be oriented and let $\alpha: I \to M$ be a differentiable curve parametrized by the arc length s with $\alpha'(s) \neq 0$, $s \in I$. Let V be a parallel vector field along α and let $\varphi = \operatorname{ang}(V, \alpha'(s))$, where the angle is measured in the given orientation. Then*

$$k_g(s) = \frac{d\varphi}{ds}.$$

Proof. Choose two frames $\{e_1, e_2\}$ and $\{\bar{e}_1, \bar{e}_2\}$ around $\alpha(s)$ as follows: $e_1 = V/|V|$ and e_2 is normal to e_1 in the positive direction; $\bar{e}_1 = \alpha'(s)$, and $\bar{e}_2$ is normal to $\bar{e}_1$ in the positive direction. As usual, they are first defined along a small interval of the curve α about $\alpha(s)$, and then extended to a neighborhood of $\alpha(s)$ in M. Denote by ω_{12} and $\bar{\omega}_{12}$ the connection forms associated to $\{e_1, e_2\}$ and $\{\bar{e}_1, \bar{e}_2\}$, respectively.

Now φ is the angle from e_1 to $\bar{e}_1$; φ is only defined up to a constant, but $d\varphi$ is well defined, and

$$d\varphi = \alpha^*\bar{\omega}_{12} - \alpha^*\omega_{12}.$$

Since e_1 is a parallel field along α, $\alpha^*\omega_{12} = 0$. Also, since $\bar{e}_1 = \alpha'(s)$, we have that

$$k_g = (\alpha^*\bar{\omega}_{12})\left(\frac{d}{ds}\right) = d\varphi\left(\frac{d}{ds}\right) = \frac{d\varphi}{ds},$$

as we wished. □

The proof of the above proposition also contains the following interpretation of the Gaussian curvature in terms of parallel transport. Let $p \in M^2$, and $D \subset M$ be a neighborhood of p homeomorphic to a disk with smooth boundary ∂D. Let $q \in \partial D$ and $V_0 \in T_qM$, $|V_0| = 1$, and transport V parallelly around the closed curve ∂D. When V returns to q, it makes an angle φ with the initial position V_0. Parametrizing ∂D as $\alpha(s)$, where s is the arc length of ∂D, and using the frames $\{e_1(s) = \alpha'(s), e_2(s)\}$, $\{\bar{e}_1(s) = V(s), \bar{e}_2(s)\}$ as in the above proof, we obtain

$$-\int_{\partial D} \alpha^*(\omega_{12}) = \int_{\partial D} d\varphi = \varphi.$$

On the other hand, by Stokes theorem,

$$\varphi = -\int_{\partial D} \alpha^*(\omega_{12}) = -\int_D d\omega_{12} = \int_D K\sigma.$$

It follows, by the mean value theorem of integral calculus, that

$$K(p) = \lim_{D \to p} \frac{\varphi}{\text{area } D},$$

that is, the Gaussian curvature at p measures how different from the identity is parallel transport along small circles about p.

EXERCISES

1) *(The flat torus).* Let $f: \mathbf{R}^2 \to \mathbf{R}^4$ be given by

$$f(x, y) = (\cos x, \sin x, \cos y, \sin y), \quad (x, y) \in \mathbf{R}^2.$$

Prove that:
a) f is an immersion and $f(\mathbf{R}^2)$ is homeomorphic to a torus,
b) The frame $e_1 = \frac{\partial f}{\partial x}$, $e_2 = \frac{\partial f}{\partial y}$ in $f(\mathbf{R}^2) \subset \mathbf{R}^4$ is orthonormal in the metric of $f(\mathbf{R}^2)$ induced by $\mathbf{R}^4$. Compute ω_1, ω_2, ω_{12},
c) The Gaussian curvature of the induced metric is identically zero.

2) *(The hyperbolic plane).* Let H^2 be the upper half-plane, that is,

$$H^2 = \{(x, y) \in \mathbf{R}^2; y > 0\}$$

Consider in H^2 the following inner product: If $(x, y) \in H^2$ and $u, v \in T_pH^2$, then

$$\langle u, v \rangle_p = \frac{u \cdot v}{y^2},$$

where $u \cdot v$ is the canonical inner product of $\mathbf{R}^2$. Prove that this is a Riemannian metric in H^2 whose Gaussian curvature is $K \equiv -1$; with this Riemannian metric H^2 is called the hyperbolic plane.
Hint: Choose the orthonormal frame $e_1 = \frac{a_1}{y}$, $e_2 = \frac{a_2}{y}$, where $\{a_1, a_2\}$ is the canonical frame of $\mathbf{R}^2$.

3) Let M^2 be a Riemannian manifold of dimension two. Let $f: U \subset \mathbf{R}^2 \to M$ be a parametrization of M^2 such that $f_u = df(\frac{\partial}{\partial u})$ and $f_v = df(\frac{\partial}{\partial v})$, $(u, v) \in U$, are orthogonal. Set $E = \langle f_u, f_u \rangle$ and $G = \langle f_v, f_v \rangle$. Choose an orthonormal frame $e_1 = f_u/\sqrt{E}$, $e_2 = f_v/\sqrt{G}$ in U. Show that:
a) The associated coframe is given by

$$\omega_1 = \sqrt{E}du, \quad \omega_2 = \sqrt{G}dv.$$

b) The connection form is given by

$$\omega_{12} = -\frac{(\sqrt{E})_v}{\sqrt{G}}du + \frac{(\sqrt{G})_u}{\sqrt{E}}dv.$$

Hint: Use the fact that $\omega_{12}(e_i) = d\omega_i(e_1, e_2)$, $i = 1, 2$.

c) The Gaussian curvature of M^2 is

$$K = -\frac{1}{\sqrt{EG}}\left\{\left(\frac{(\sqrt{E})_v}{\sqrt{G}}\right)_v + \left(\frac{(\sqrt{G})_u}{\sqrt{E}}\right)_u\right\}.$$

4) Let $S^2 = \{(x,y,z) \in \mathbf{R}^3; x^2+y^2+z^2=1\}$. Prove that there exists no differentiable nonzero vector field X on S^2.
Hint: Assume the existence of such a field X. Let $e_1 = X/|X|$ e consider the orthonormal oriented frame $\{e_1.e_2\}$. Then $d\omega_{12} = -K\omega_1 \wedge \omega_2 = -\sigma$, hence

$$\text{area } S^2 = \int_{S^2} \sigma = -\int_{S^2} d\omega_{12} = -\int_{\partial S^2} \omega_{12} = 0,$$

which is a contradiction.

5) Consider $\mathbf{R}^2$ with the following inner product: If $p = (x,y) \in \mathbf{R}^2$ and $u, v \in T_p\mathbf{R}^2$, then

$$\langle u, v\rangle_p = \frac{u \cdot v}{(g(p))^2},$$

where $u \cdot v$ is the canonical inner product of $\mathbf{R}^2$ and $g: \mathbf{R}^2 \to \mathbf{R}$ is a differentiable positive function. Prove that the Gaussian curvature of this metric is

$$K = g(g_{xx} + g_{yy}) - (g_x^2 + g_y^2).$$

6) Let $M^2 \subset \mathbf{R}^3$ be a surface with the induced metric. Let $p \in M^2$, $x \in T_pM^2$ and Y be a vector field tangent to M^2. Show that

$$(\nabla_x Y)(p) = \text{projection onto } T_pM \text{ of } \left(\frac{dY(\alpha(s))}{ds}\right)(0),$$

where $\alpha: I \to M$ is a differentiable curve, $s \in I$, and $\frac{dY}{ds}$ is the usual derivative of vectors in $\mathbf{R}^3$. Conclude that a curve $\gamma(s)$ in M, parametrized by the arc length s, is a geodesic in M if and only if the "acceleration" vector $\frac{d^2\gamma}{ds^2}$ in $\mathbf{R}^3$ is everywhere perpendicular to M.

7) Let $S^2 = \{(x,y,z) \in \mathbf{R}^3; x^2+y^2+z^2=1\}$ be the unit sphere with the metric induced from $\mathbf{R}^3$. Show that:
   a) The geodesics of S^2 are its great circles,
   b) The antipodal map $A: S^2 \to S^2$ given by $A(x,y,z) = (-x,-y,-z)$ is an isometry,
   c) The projective plane $P^2(\mathbf{R})$ (cf. Example 7 of Chapter 2) can be given a Riemannian metric such that the canonical projection $\pi: S^2 \to P^2(\mathbf{R})$ is a local isometry (that is, each $p \in S^2$ has a neighborhood V such that the restriction π/V is an isometry).

8) Let M^2 be a Riemannian manifold (of dimension two). The goal of the exercise is to show that the Gaussian curvature K of M is identically zero if and only if M is locally euclidean, that is, there exist local coordinates

(u, v) around any point such that the first fundamental form $I = du^2 + dv^2$. Clearly if I is as above, $K = 0$. To prove the converse, proceed as follows:

a) Choose a frame $\{e_1, e_2\}$ around $p \in M$. Since $d\omega_{12} = -K\omega_1 \wedge \omega_2 = 0$, by Poincare's Lemma, there exists a function θ defined in a neighborhood V of such that $d\theta = \omega_{12}$,

b) Choose another frame $\{\bar{e}_1, \bar{e}_2\}$ by setting ang $(e_1, \bar{e}_1) = \theta$. Show that the connection form $\bar{\omega}_{12}$ of this frame vanishes identically,

c) Show that $\bar{\omega}_{12} = 0$ implies that $d\bar{\omega}_1 = d\bar{\omega}_2 = 0$ and use again Poincare's Lemma to obtain the required local coordinates.

6. The Theorem of Gauss-Bonnet and the Theorem of Morse

1. The Theorem of Gauss-Bonnet

The considerations of the last chapter were strictly local. However, one of the most interesting features of differential geometry is the connection between local properties and properties that depend on the entire surface. One of the most striking of such properties is the so-called Gauss-Bonnet theorem which we intend to prove in this section.

In his fundamental work (Considerations on curved surfaces, 1827), Gauss proved the special case of this theorem for geodesic triangles and foresaw its importance for the development of differential geometry. The theorem for more general regions is due to O. Bonnet (Jour. Ecole Polytech. 19 (1848), 1-146). With the advent of Topology, it became soon clear that a global formulation of the Gauss-Bonnet theorem would be an important link between Geometry and Topology. The extension of this result to higher dimensions became then an important mathematical problem. After some preliminary work by Allendoerfer and Weil, a satisfactory solution was obtained in 1944 by S.S. Chern, as an application of the method of moving frames. We will come back to that in Remark 2 of this section.

Before starting, we want to make the general remark that any differentiable manifold M^n (Hausdorff and with countable basis) can be given a Riemannian metric. The proof depends on the existence of a partition of unity. For the compact case (which is the only one we will use), it suffices to define arbitrarily an inner product $\langle , \rangle^\alpha$ on each coordinate neighborhood $f^\alpha(U^\alpha)$ of a finite differentiable structure of M^n, and to set

$$\langle , \rangle_p = \sum_\alpha \varphi_\alpha(p) \langle , \rangle_p^\alpha, \quad p \in M^n,$$

where φ_α is a differentiable partition of unit subordinate to the (finite) covering $f_\alpha(U_\alpha)$.

From now on, M will denote a compact, oriented, differentiable manifold *of dimension two*. Let X be a differentiable vector field on M. A point $p \in M$ is a *singular point* of X if $X(p) = 0$; the singular point p is *isolated* if there exists a neighborhood $V \subset M$ of p which contains no singular point other than p. In what follows, it will be convenient to choose V homeomorphic to an open disk in the plane. Notice that the number of isolated singular points is finite, since M is compact.

To each isolated singular point of X, we are going to associate an integer to be called the index of X at p, as follows. First, choose a Riemannian metric on M, and consider the moving frame $\{\bar{e}_1, \bar{e}_2\}$, where $\bar{e}_1 = X/|X|$ and $\bar{e}_2$ is a unit vector field orthogonal to $\bar{e}_1$ and in the orientation of M. This determines differential forms $\overline{\omega}_1, \overline{\omega}_2, \overline{\omega}_{12}$ in $V - \{p\}$. Next, we choose another moving frame $\{e_1, e_2\}$, in the same orientation as before, defined throughout V, thus obtaining forms $\omega_1, \omega_2, \omega_{12}$ in V. The difference

$$\overline{\omega}_{12} - \omega_{12} = \tau$$

is defined in $V - \{p\}$.

Now consider a simple closed curve C that bounds a compact region of V containing p in its interior; C will be oriented as the boundary of this region. By Lemma 5 of Chapter 5, the restriction of τ to C is the differential of the angle $\varphi(t)$ between e_1 and $\bar{e}_1$ along C. Thus

$$\int_C \tau = \int_C d\varphi = 2\pi I.$$

The integer I is called the *index of X at p*.

Notice that in the definition of index we made various choices, namely, the choice of a Riemannian metric, the choice of a frame $\{e_1, e_2\}$, and the choice of a curve C. It is clearly necessary to show that the index I does not depend on these choices.

Before that, let us look at some examples of singularities of vector fields in the plane. Intuitively, the index is the number of "turns" given by the vector field as we go along a simple closed curve around an isolated singularity. The figure below (Fig. 6.1) displays some vector fields in the plane (described by their trajectories) with isolated singularities, and their respective indices.

Lemma 1. *The definition of I does not depend on the curve C.*

Proof. Let C_1 and C_2 be two simple closed curves around p, as in the definition of index. Assume first that C_1 and C_2 do not intersect and consider the annular region Δ bounded by C_1 and C_2. Let I_1 be the index computed with C_1 and I_2 be the index computed with C_2. By Stokes theorem, and the fact that $d\tau = 0$,

$$I_1 - I_2 = \frac{1}{2\pi}\int_{C_1} \tau - \frac{1}{2\pi}\int_{C_2} \tau = \frac{1}{2\pi}\int_{\Delta} d\tau = 0,$$

and this proves the Lemma in this case. If C_1 and C_2 intersect, we choose a curve C_3 that does not intersect both C_1 and C_2. By applying the above, we conclude that $I_1 = I_3 = I_2$. □

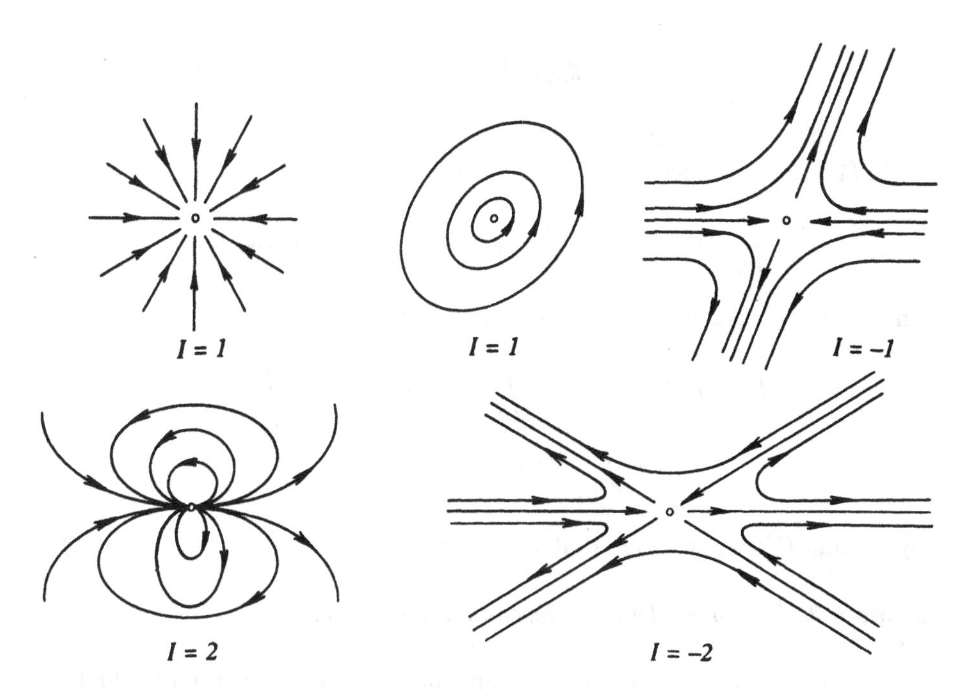

Fig. 6.1

Lemma 2. *The definition of I does not depend on the choice of the frame $\{e_1, e_2\}$. More precisely, let $S_r = \partial B_r$ be the boundary of a disk of radius r and center p, and consider the frame $\{\bar{e}_1, \bar{e}_2\}$ of the definition. Then, the limit*

$$\lim_{r \to 0} \frac{1}{2\pi} \int_{S_r} \overline{\omega}_{12} = \bar{I}$$

exists, and $\bar{I} = I$.

Proof. Let S_{r_1}, S_{r_2} be concentric circles, $r_2 < r_1$, and let $\triangle$ be the annular region bounded by S_{r_1} and S_{r_2}. By Stokes theorem,

$$\int_{S_{r_1}} \overline{\omega}_{12} - \int_{S_{r_2}} \overline{\omega}_{12} = \int_{\triangle} d\overline{\omega}_{12} \to 0, \quad \text{as } r_1, r_2 \to 0. \tag{1}$$

Notice that $\overline{\omega}_{12}$ is not defined in B_{r_2}; however, $d\overline{\omega}_{12} = -K\sigma$ is certainly defined everywhere. It follows that any sequence

$$\int_{S_{r_1}} \overline{\omega}_{12}, \ldots, \int_{S_{r_n}} \overline{\omega}_{12}, \ldots,$$

with $\{r_n\} \to 0$, is a Cauchy sequence, hence converges. Thus the limit

$$\lim_{r\to 0} \frac{1}{2\pi} \int_{S_r} \overline{\omega}_{12} = \bar{I}$$

exists, and we want to show that $\bar{I} = I$.

In (1), fix r_1 and let r_2 go to zero. Then

$$\int_{S_{r_1}} \overline{\omega}_{12} - 2\pi\bar{I} = \int_{B_{r_1}} d\overline{\omega}_{12} = -\int_{B_{r_1}} K\overline{\omega}_1 \wedge \overline{\omega}_2. \tag{2}$$

On the other hand, since $\overline{\omega}_{12} = \omega_{12} + \tau$, we have

$$\begin{aligned} \int_{S_{r_1}} \overline{\omega}_{12} &= \int_{S_{r_1}} \omega_{12} + \int_{S_{r_1}} \tau = \int_{B_{r_1}} d\omega_{12} + \int_{S_{r_1}} \tau \\ &= -\int_{B_{r_1}} K\omega_1 \wedge \omega_2 + 2\pi I. \end{aligned} \tag{3}$$

By (2) and (3), we conclude that $I = \bar{I}$, as we wished. □

Lemma 3. *The index does not depend on the metric.*

Proof. Let $\langle\,,\rangle_0$ and $\langle\,,\rangle_1$ be two Riemannian metrics on M. Let, for $t \in [0,1]$,

$$\langle\,,\rangle_t = t\langle\,,\rangle_1 + (1-t)\langle\,,\rangle_0.$$

Then $\langle\,,\rangle_t$ is easily seen be a positive definite inner product on M which varies differentiably with p. Thus $\langle\,,\rangle_t$ is a one-parameter family of metrics on M that starts with $\langle\,,\rangle_0$ and ends with $\langle\,,\rangle_1$. Let I_0, I_t and I_1 be the corresponding indices. It is easily checked, by using Lemmas 1 and 2, that I_t is a continuous function of t. Being an integer, $I_t =$ const. , $t \in [0,1]$. Thus $I_0 = I_1$, as we wished. □

We can now state and prove the Gauss-Bonnet Theorem in the following form.

Theorem 1. *Let M^2 be a two-dimensional compact oriented differentiable manifold. Let X be a differentiable vector field on M with isolated singularities $p_1, \dots, p_k$ whose indices are $I_1, \dots, I_k$. Then, for any Riemannian metric on M,*

$$\int_M K\sigma = 2\pi \sum_{i=1}^{k} I_i,$$

where K is the Gaussian curvature of the metric and σ is its element of area.

Proof. Consider in $M^2 - \underset{i}{U}\{p_i\}$ the frame $\{\bar{e}_1 = X/|X|,\ \bar{e}_2\}$, where $\bar{e}_2$ is a unit vector field orthogonal to $\bar{e}_1$ in the orientation of M. Let us denote by

B_i a ball with center p_i which is such that it contains no singular point other than p_i. From Stokes theorem, we have that

$$\int_{M-\underset{i}{U}B_i} K\bar{\omega}_1 \wedge \bar{\omega}_2 = -\int_{M-\underset{i}{U}B_i} d\bar{\omega}_{12} = \int_{U(\partial B_i)} \bar{\omega}_{12} = \sum_i \int_{\partial B_i} \bar{\omega}_{12},$$

where ∂B_i has the orientation induced by B_i (this is the opposite of the orientation of $M - B_i$, hence the change of sign in the second equality). Now, take the limit of the above, as the radii of B_i go to zero, and use Lemma 2 to obtain that

$$\int_M K\omega_1 \wedge \omega_2 = 2\pi \sum_i I_i,$$

as we wished. □

Notice that the right hand side does not depend on the vector field X and the left hand side does not depend on the metric. Thus we obtain the striking conclusion that $\sum I_i$ is the same for all vector fields with isolated singularities, and $\int_M K\sigma$ is the same for all Riemannian metrics on M.

The number $\sum_{i=1}^{k} I_i$ is called the *Euler-Poincaré characteristic* of M and is also denoted by $\chi(M)$. By the above, $\chi(M)$ is invariant by diffeomorphisms and $\frac{1}{2\pi}\int_M K\sigma$ does not depend on the metric on M and it is equal to this invariant.

Remark 1. Another way of introducing $\chi(M)$, for a compact surface M, is to decompose M into a finite number of curvilinear triangles in such a way that the intersection of two such triangles be either empty, a common edge, or a common vertex of the triangles (such a decomposition is called a *triangulation of* M and it is possible to prove that it always exists.)

Let us denote by F the total number of triangles, V the total number of vertices and A the total number of edges of such a triangulation. By definition

$$\chi(M) = V - A + F.$$

We will show that, when M is orientable, this definition agrees with the previous one. For that, choose a triangulation, and consider the vector field in M indicated by its trajectories in Fig. 6.2 below (actually we only display the field in two triangles of the triangulation). The indices of the vector fields at the points B, C and D are, respectively, 1, 1 and -1. Thus the total sum of the indices is

$$\sum_i I_i = V - A + F.$$

Since $\sum_i I_i = \chi(M)$ does not depend on the chosen field, we conclude that $V - A + F = \chi(S)$.

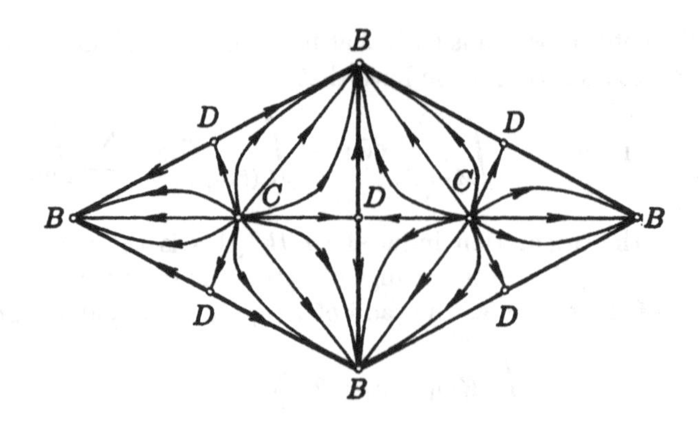

Fig. 6.2

Example 1. Let us compute $\chi(M^2)$ for the sphere S^2 and the torus T.

For the sphere, we choose the metric induced by $\mathbf{R}^3$ on $S^2 = \{p \in \mathbf{R}^3; |p| = 1\}$ for which $K \equiv 1$ (cf. Example 1 of Chap. 4). Thus $\int_S K\sigma =$ area $S^2 = 4\pi$, hence $\chi(S^2) = 2$. As a consequence, we see that every tangent vector field on S^2 has at least one singularity.

For the torus T, we know that it is possible to introduce in T a metric with $K \equiv 0$ (cf. Exercise 1 in Chap. 4). Since $\chi(T)$ does not depend on the metric, $\chi(T) = 0$.

Remark 2. The proof of the Gauss-Bonnet Theorem presented here is essentially due to S.S. Chern. The crucial point of the proof is the existence of a form $\overline{\omega}_{12}$ in $M - U\{p_i\}$ that depends on the vector field X but whose differential $d\overline{\omega}_{12} = -K\sigma$ does not depend on X and is globally defined on M. The idea that this was the crucial part of the proof lead Chern to give a proof of a generalized Gauss-Bonnet theorem [CHER]. Furthermore, such an idea opened the way for other Gauss-Bonnet type theorems and is in the basis of the creation of the so-called Chern classes.

Remark 3. It can be shown that a surface M (connected, compact and orientable) is homeomorphic to M' if and only if $\chi(M) = \chi(M')$. A proof can be found in W. Massey, [MAS].

We now go into the Gauss-Bonnet Theorem for surfaces with boundary.

Theorem 2. *Let M be an oriented, compact, two-dimensional differentiable manifold with boundary ∂M, and let X be a differentiable vector field on M such that it is transversal to ∂M (that is, X is nowhere tangent to ∂M). Assume that the singularities $p_1, \ldots, p_k$ of X are isolated, do not belong to*

∂M and denote their indices by $I_1, \dots, I_k$. Then, for any Riemannian metric on M,

$$\int_M K\sigma + \int_{\partial M} k_g ds = 2\pi \sum I_i,$$

where k_g is the geodesic curvature of ∂M and ds is the arc element of ∂M.

Proof. Choose a Riemannian metric on M and consider in M the orthonormal oriented frame $\bar{e}_1 = X/|X|, \bar{e}_2$. Choose, in a neighborhood $V \subset M$ of ∂M, another oriented frame $\{e_1, e_2\}$ such that, restricted to ∂M, e_1 is tangent to ∂M. Then

$$i^*\bar{\omega}_{12} = i^*\omega_{12} + d\varphi,$$

where $i: \partial M \to M$ is the inclusion map and φ is the angle between $\bar{e}_1$ and e_1 along ∂M.

Let B_i be a ball of center p_i, $i = 1, \dots, k$, so that B_i contains no singular point other than p_i. Then

$$\int_{M-UB_i} K\bar{\omega}_1 \wedge \bar{\omega}_2 = -\int_{M-UB_i} d\bar{\omega}_{12} = \int_{U\partial B_i} \bar{\omega}_{12} - \int_{\partial M} i^*\bar{\omega}_{12},$$

hence

$$\int_{M-UB_i} K\bar{\omega}_1 \wedge \bar{\omega}_2 + \int_{\partial M} i^*\bar{\omega}_{12} = \sum_{i=1}^{k} \int_{\partial B_i} \bar{\omega}_{12}.$$

But, by the definition of geodesic curvature,

$$\int_{\partial M} i^*\bar{\omega}_{12} = \int_{\partial M} i^*\omega_{12} + \int_{\partial M} d\varphi = \int_{\partial M} k_g ds + \int_{\partial M} d\varphi.$$

Since $\bar{e}_1 = X/|X|$ is nowhere tangent to ∂M, $\int_{\partial M} d\varphi = 0$. Therefore, by taking the limit when the radius of B_i goes to zero, we obtain

$$\int_M K\sigma + \int_{\partial M} k_g ds = 2\pi \sum I_i,$$

as we wished. □

The number $\sum I_i$ is again an invariant by diffeomorphisms of the (compact, oriented) surface M with boundary ∂M, and is again called the Euler-Poincaré characteristic $\chi(M)$ of M. Let us compute some examples:

Example 2. Consider the segment of a right circular cylinder M bounded by two parallel circles. The metric induced by $\mathbf{R}^3$ on M has vanishing Gaussian curvature and the two boundary circles that constitute ∂M are geodesics in M. Thus

$$\chi(M) = \sum I_i = \frac{1}{2\pi}\left(\int_M K\sigma + \int_{\partial M} k_g ds\right) = 0.$$

Example 3. The Euler-Poincaré characteristic of a plane disk D of radius r is equal to one. In fact, taking the canonical metric in the plane of the disk, we obtain that $K \equiv 0$ and that the geodesic curvature of ∂D is $k_g = 1/r$. Therefore,

$$\chi(D) = \sum I_i = \frac{1}{2\pi} \int_{\partial D} k_g ds = \frac{1}{2\pi} \frac{1}{r} 2\pi r = 1.$$

Remark 4. The Gauss-Bonnet Theorem still holds for compact surfaces with boundaries and corners, that is, those surfaces for which the boundary fails to be regular at finitely many points which are called *corners*. Each corner q_j, $j = 1, \ldots, n$, gives rise to an *external* angle α_j (which is the positive angle made by the tangents at the corner) and this must be added to the total geodesic curvature, so that the theorem now reads:

$$\int_M K\sigma + \int_{\partial M} k_g ds + \sum_{j=1}^{n} \alpha_j = 2\pi \sum_{i=1}^{k} I_i.$$

The proof involves a certain number of technicalities and we will not enter into that here.

2. The Theorem of Morse

Closely related with the Gauss-Bonnet theorem is a relation, due to Marston Morse, between critical points of a certain class of functions on a compact surface M^2 and the topology of M^2. In this section we intend to establish this relation.

The class of functions that we have in mind can be described as follows. As always, M^2 will denote an oriented compact differentiable manifold of dimension two.

Let $f: M^2 \to \mathbf{R}$ be a differentiable function on M^2. The point $p \in M$ is a *critical point* of f if $df_p = 0$. If we choose any metric on M^2 and define a vector field grad f by

$$\langle \operatorname{grad} f(p), v \rangle = df_p(v), \text{ for any } v \in T_pM^2,$$

we obtain that p is a critical point of f if and only if, for any metric on M^2, p is a singular point of grad f.

A critical point p of $f: M^2 \to \mathbf{R}$ is said to be *nondegenerate* if for some parametrization $g: U \subset \mathbf{R}^2 \to M^2$ around $p = g(0,0)$, we have that $\det(A) \neq 0$, where A is the matrix

$$A = \begin{pmatrix} \frac{\partial^2 (f \circ g)}{\partial x^2} & \frac{\partial^2 (f \circ g)}{\partial x \partial y} \\ \frac{\partial^2 (f \circ g)}{\partial y \partial x} & \frac{\partial^2 (f \circ g)}{\partial y^2} \end{pmatrix} (0,0), \qquad (x,y) \in U \subset \mathbf{R}^2.$$

Since at a critical point the first derivatives vanish, it is easily checked that the fact that $\det A \neq 0$ does not depend on the parametrization g.

Nondegenerate critical points are the simplest type of critical points. The behavior of the function $f \circ g = h$ in a neighborhood of such a point can be easily described using Taylor's expansion:

$$d = h(x,y) - h(0,0) = \frac{1}{2}\left\{\left(\frac{\partial^2 h}{\partial x^2}\right)_0 x^2 + 2\left(\frac{\partial^2 f}{\partial x \partial y}\right)_0 xy + \left(\frac{\partial^2 f}{\partial y^2}\right)_0 y^2\right\}$$
$$+ \text{ terms of higher order }.$$

In fact, if $x^2 + y^2$ is small enough, and $\det A \neq 0$, the sign of d is controled by the sign of the quadratic form in the right hand side. Thus if $\det A > 0$, we have one of the following alternatives:

a) $d < 0$ in some neighborhood of p; p is then called a *point of maximum* for f.

b) $d > 0$ in some neighborhood of p; p is then a *point of minimum* for f.

On the other hand, if $\det A < 0$, there exist exactly two distinct directions for which the quadratic form vanishes; for all other directions, either d is positive (a direction of minimum) or negative (a direction of maximum). Such a point is called a *saddle point*.

Now we can state the relation we want to prove

Theorem 3. (Morse) *Let $f: M^2 \to \mathbf{R}$ be a differentiable function on a compact oriented surface M^2 such that all its critical points are nondegenerate. Let us denote by M, m, and s the number of points of maximum, minimum and saddle, respectively, of f. Then $M - s + m$ does not depend on f; more precisely,*

$$M - s + m = \chi(M^2).$$

Before going into the proof, we need some preliminary considerations. From now on, assume that we have chosen a Riemannian metric on M^2. As we have seen, p is a critical point of f if and only if the vector field $\operatorname{grad} f$ has a singularity at p. How does the fact that p is a nondegenerate critical point reflects on the field $\operatorname{grad} f$?

To answer that question, we will need a definition.

Let X be a differentiable vector field on M^2, p be a singular point of X and $g = U \subset \mathbf{R}^2 \to M$ a parametrization around $p = g(0,,0)$. In the basis $\{\frac{\partial}{\partial x}, \frac{\partial}{\partial y}\}, (x,y) \in U$, associated to the parametrization g, we can write X as

$$X = \alpha(x,y)\frac{\partial}{\partial x} + \beta(x,y)\frac{\partial}{\partial y},$$

where α and β are differentiable functions in U, and, since p is a singular point, $\alpha(0,0) = \beta(0,0) = 0$. Let A_g be the matrix of the *linear part* of X, that is,

$$A_g = \begin{pmatrix} \left(\frac{\partial\alpha}{\partial x}\right)_0 & \left(\frac{\partial\alpha}{\partial y}\right)_0 \\ \left(\frac{\partial\beta}{\partial x}\right)_0 & \left(\frac{\partial\beta}{\partial y}\right)_0 \end{pmatrix}$$

We say that p is a *simple singularity* of X if $\det A_g \neq 0$. This definition does not depend on the parametrization g, since for another parametrization $\tilde{g}$, we easily obtain

$$A_{\tilde{g}} = dh \circ A_g \circ (dh)^{-1},$$

where h is the change of coordinates.

We can now answer our question.

Proposition 1. *Let $p \in M^2$ be a critical point of a differentiable function $f: M^2 \to \mathbf{R}$ on a Riemannian manifold M^2. Then p is a nondegenerate critical point of f if and only if p is a simple singularity of* grad f.

Proof. Let $g: U \subset \mathbf{R}^2 \to M^2$ be a parametrization around $p = g(0,0)$ such that $\langle \frac{\partial}{\partial x}, \frac{\partial}{\partial y} \rangle = 0$, for all $(x,y) \in U$; such a parametrization always exist (cf.[**dC**], §3.4). Let us compute grad f in this parametrization.

Let $\operatorname{grad} f = \alpha \frac{\partial}{\partial x} + \beta \frac{\partial}{\partial y}$, and let $Y = y_1 \frac{\partial}{\partial x} + y_2 \frac{\partial}{\partial y}$ be an arbitrary vector of $T_{g(x,y)}M^2$. By definition,

$$\langle \operatorname{grad} f, Y \rangle = df(Y),$$

hence

$$\alpha y_1 \langle \frac{\partial}{\partial x}, \frac{\partial}{\partial x} \rangle + \beta y_2 \langle \frac{\partial}{\partial y}, \frac{\partial}{\partial y} \rangle = \frac{\partial (f \circ g)}{\partial x} y_1 + \frac{\partial (f \circ g)}{\partial y} y_2,$$

for all pairs (y_1, y_2). Thus setting $\langle \frac{\partial}{\partial x}, \frac{\partial}{\partial x} \rangle = g_{11}$ and $\langle \frac{\partial}{\partial y}, \frac{\partial}{\partial y} \rangle = g_{22}$, we obtain

$$\operatorname{grad} f = \frac{\partial (f \circ g)}{\partial x} \frac{1}{g_{11}} \frac{\partial}{\partial x} + \frac{\partial (f \circ g)}{\partial y} \frac{1}{g_{22}} \frac{\partial}{\partial y}.$$

Observe now that it is possible to choose the parametrization g so that, at p, $g_{11}(p) = g_{22}(p) = 1$. Thus the linear part of grad f is given, in this parametrization, by

$$A_g = \begin{pmatrix} \frac{\partial^2 (f \circ g)}{\partial x^2} & \frac{\partial^2 (f \circ g)}{\partial x \partial y} \\ \frac{\partial^2 (f \circ g)}{\partial x \partial y} & \frac{\partial^2 (f \circ g)}{\partial y^2} \end{pmatrix} (0,0).$$

It now suffices to observe that both conditions in the statement of the Proposition are equivalent to $\det(A_g) \neq 0$. □

It turns out that simple singularities of vector fields are all isolated.

Lemma 4. *Let $p \in M^2$ be a simple singularity of a differentiable vector field X on M. Then p is an isolated singular point of X.*

Proof. Let $g: U \subset \mathbf{R}^2 \to M^2$ be a parametrization around $p = g(0,0)$, and let

$$X = \alpha \frac{\partial}{\partial x} + \beta \frac{\partial}{\partial y}$$

be the expression of X in this parametrization.

Consider the map $\varphi: U \subset \mathbf{R}^2 \to \mathbf{R}^2$ given by $\varphi(x,y) = (\alpha(x,y), \beta(x,y))$. Since p is a simple singularity,

$$\det(d\varphi_0) = \det A_g \neq 0, \quad \text{at } p.$$

By the inverse function theorem, there exists a neighborhood $V \subset U$ of $(0,0)$ where φ is bijective. Thus if $\alpha(x,y) = \beta(x,y) = 0$, $(x,y) \in V$, then $x = y = 0$, that is, in $g(V)$ there is no singular point other than p. □

Corollary. *Nondegenerate critical points of a differentiable functions $f = M^2 \to \mathbf{R}$ are isolated.*

From Lemma 4 it follows that it makes sense to talk about the index of a simple singularity. We will now show that they are easily computable.

Proposition 2. *Let $p \in M^2$ be a simple singular point of a differentiable vector field X on M^2. Then the index of X at p is either $+1$ (if the determinant of the linear part of X is positive) or -1 (if the determinant of the linear part of X is negative).*

Proof. Since the index is local and does not depend on the choice of a metric, we can take $M^2 = \mathbf{R}^2$ with the canonical metric, and assume that $p = (0,0) = 0$. Thus $X: \mathbf{R}^2 \to \mathbf{R}^2$ is a differentiable map with $X(0) = 0$, and the linear part (the differential) of X at 0 is given by

$$dX_0: \mathbf{R}^2 \to \mathbf{R}^2, \quad dX_0(p) = \lim_{t \to 0} \frac{X(tp)}{t}, \qquad p \in \mathbf{R}^2.$$

Let us define a map $F: \mathbf{R}^2 \times I \to \mathbf{R}^2$ by

$$F(p,t) = \begin{cases} \frac{X(tp)}{t}, & \text{if } t \neq 0 \\ dX_0(p), & \text{if } t = 0 \end{cases} \qquad (p,t) \in \mathbf{R}^2 \times I.$$

We claim that F is continuous. To see that, we need the following fact from Calculus: *If $f: \mathbf{R}^n \to \mathbf{R}$ is differentiable and* $f(0,0,\dots,0) = 0$, we can write

$$f(x_1,\dots,x_n) = \int_0^1 \frac{df(tx_1,\dots,tx_n)}{dt}\,dt$$
$$= \sum_{i=1}^{n} x_i \int_0^1 \frac{\partial f}{\partial x_i}(tx_1,\dots,tx_n)dt;$$

then, *by setting*,

$$h_i(x_1,\dots,x_n) = \int_0^1 \frac{\partial f}{\partial x_i}(tx_1,\dots,tx_n)dt$$

we obtain that h_i is differentiable, $h_i(0,\dots,0) = \frac{\partial f}{\partial x_i}(0,\dots,0)$, and

$$f(x_1,\dots,x_n) = \sum_i x_i h_i(x_1,\dots,x_n).$$

It follows, by setting

$$X(x_1,x_2) = (\alpha(x_1,x_2),\ \beta(x_1,x_2)), \qquad (x_1,x_2) \in \mathbf{R}^2,$$

that we can write:

$$\alpha(x_1,x_2) = \sum_i x_i h_{1i}(x_1,x_2),$$
$$\beta(x_1,x_2) = \sum_i x_i h_{2i}(x_1,x_2), \qquad i = 1,2,$$

where h_{ij}, $i,j = 1,2$, are differentiable functions. Since, even for $t = 0$,

$$F((x_1,x_2),t) = (\sum_i x_i h_{1i}(tx_1,tx_2),\ \sum_i x_i h_{2i}(tx_1,tx_2)),$$

we conclude that F is continuous, as we claimed.

F maps continuously the vector field $dX_0 = F(p,0)$ to the vector field $X = F(p,1)$. Thus the index of X at 0 is equal to the index of dX_0 at 0, and it suffices to compute the latter.

For that, consider the circle $C = \{(x_1,x_2) \in \mathbf{R}^2;\ x_1^2 + x_2^2 = 1\}$. If $q_1, q_2 \in C$ and $q_1 \neq q_2$, since dX_0 is non-singular, $dX_0(q_1) \neq dX_0(q_2)$. If $\det(dX_0) > 0$, the index I of dX_0 is positive and, by what we have just seen, cannot be greater than one; thus $I = 1$. If $\det(dX_0) < 0$, the index I of dX_0 is negative and, by the same token, cannot be smaller than -1; Thus $I = -1$. This proves the Proposition. □

Remark. The above is just another proof of the Lemma used to prove Theorem 3 in Chapter 2.

With all these preliminaries, the proof of Morse Theorem is almost immediate.

Proof of Theorem 3. Choose a Riemannian metric on M. Since the critical points of f are nondegenerate, the singularities of grad f are isolated and simple. Thus the index of grad f is 1, at a point where f is either maximum or minimum, or the index of grad f is -1, at a saddle point of f. It follows that $M - s + m$ is equal to the sum of the indices of the singularities of grad f. By the Gauss Bonnet theorem, such a sum does not depend on the chosen metric or on the field grad f, and it is equal to $\chi(M^2)$. □

Remark. Theorem 2 is only a sample of the deep relations established by M. Morse between the topology of differentiable manifolds and the critical points of certain classes of differentiable functions. A beautiful introduction to the subject is J. Milnor [MILN].

EXERCISES

1) Compute the Euler-Poincaré characteristic of
   a) an ellipsoid,
   b) $M = \{(x,y,z) \in \mathbf{R}^3;\ x^2 + y^4 + z^6 = 1\}$.
2) Prove that there exists no Riemannian metric on a torus T such that K is nonzero and does not change sign on T.
3) Let M^2 be a connected compact orientable manifold of dimension two. Prove that the following statements are equivalent (Assume that if $\chi(M^2) = \chi(\overline{M}^2)$ then M^2 is homeomorphic to $\overline{M}^2$):
   a) These exists a nowhere zero differentiable vector field on M^2.
   b) $\chi(M^2) = 0$.
   c) M^2 is homeomorphic to a torus.
4) Let $M^2 \subset \mathbf{R}^3$ be a regular surface in $\mathbf{R}^3$. Assume that M^2 is compact, oriented and not homeomorphic to a sphere. Show that there exist points in M^2 for which the Gaussian curvature is positive, negative and zero.
5) Let M^2 be a connected, compact, oriented Riemannian manifold of dimension two such that the Gaussian curvature K is always positive. Prove that two simple closed geodesics in M^2 have a common point.
6) Let $f\colon \mathbf{R}^2 \to \mathbf{R}$ be given by (the monkey saddle) $f(x,y) = x^3 - 3xy^2$. Let $p = (0,0) \in \mathbf{R}^2$. Show that:
   a) p is an isolated critical point of f.
   b) p is a degenerate critical point.
   c) The index of grad f at p is equal to -2.
7) Let $x\colon M^2 \to \mathbf{R}^3$ be an immersion of a two-dimensional differentiable manifold M^2 into $\mathbf{R}^3$(i.e., a surface in $\mathbf{R}^3$), and let $h_\nu\colon M \to \mathbf{R}$ be the *height function*, $h_\nu(p) = \langle x(p), \nu\rangle, p \in M$, of x relative to a fixed unit vector $\nu \in \mathbf{R}^3$($h_\nu(p)$ measures the "height" of $x(p)$ relative to a plane through the origin and perpendicular to ν.).

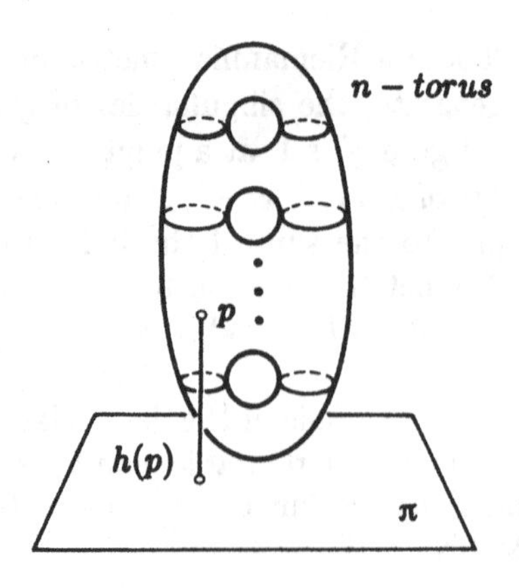

Fig. 6.3

a) Show that $p \in M$ is a critical point of h_ν if and only if $T_pM \perp \nu$.
b) Show that a critical point p of h_ν is nondegenerate if and only the *Gaussian curvature* $K(p)$ of the surface at p is nonzero.
Hint. Choose a parametrization $g: U \subset \mathbf{R}^2 \to M^2$ around p so that $g(0,0) = p$. The condition that p is nondezenerate is given by $\det A \neq 0$, where A is the matrix $\left(\frac{\partial^2(hog)}{\partial x_i \partial x_j}(0)\right), i, j = 1, 2$. Since

$$\frac{\partial^2(hog)}{\partial x_i \partial x_j}(0) = \left\langle \frac{\partial^2(xog)}{\partial x_i \partial x_j}(0), \nu(p) \right\rangle,$$

A is the matrix of the second fundamental of the surface at p in the parametrization g. But then $\det A = \pm K(p)$
c) For this part assume the following version of Sard's theorem. Let $f: M^n \to N^n$ be a differentiable map and let $q \in N$; we call q a *regular value* of f if at all points $p \in f^{-1}(q)$ the differential df_p is nonsingular (notice that if $q \notin f(M)$, then q is a regular value of f). Sard's theorem asserts that *the set of regular values of f is an open and dense subset of N*. Use Sard's theorem and part (b) to show that then exists an open and dense set U of the unit sphere of $\mathbf{R}^3$ such that if $\nu \in U$, all critical points of h_ν are nondegenerate

8) The n-torus (a torus with n holes) is a compact orientable surface M diffeomorphic to the surface that appears in Fig. 6.3 Consider the height function h_ν of M (see Exercise 7) and choose ν so that all critical points of h_ν are nondegenerate (Exercise 7(e)). Use Morse's theorem to show that the Euler characteristic of the n-torus is $2 - 2n$.
9) Let $M^2 \subset \mathbf{R}^3$ be a regular surface in $\mathbf{R}^3$, let $q \in \mathbf{R}^3$, $q \notin M^2$, and let $f: M^2 \to \mathbf{R}$ be the function given by $f_q(p)$ = distance from p to q, $p \in M^2$. Show that:

a) f_q is differentiable
b) $p \in M^2$ is a critical point of f_q if and only the straight line $\overline{pq}$ is normal to M^2.
c) The critical point p is degenerate if and only $f_q(p) = 1/k_i, i = 1, 2$, where k_1 and k_2 are the principal curvatures of M^2 at p relative to the normal $\vec{pq}$

References

[CHER] Chern, S.S., A simple intrinsic proof of the Gauss-Bonnet formula for closed Riemannian manifolds, Annals of Math. 45(1944), 747-752.

[dC] do Carmo, M., Differential Geometry of Curves and Surfaces, Prentice-Hall, 1976.

[FIG] Figueiredo, D., A simplified proof of the divergence theorem, American Math. Monthly, 71(1964), 619-622.

[FISC] Fischer, G., Mathematical Models, Vieweg, Wiesbaden, 1986.

[HIR] Hirsh, M., Differential Topology, Springer-Verlag, Berlin, 1976.

[KELL] Kellog, O., Foundations of Potential Theory, Dover Publications, New York, 1954.

[LIM 1] Lima, E., Curso de Análise, vol. 2, Projeto Euclides, IMPA, Rio de Janeiro, 2ª ed. 1985 (in Portuguese).

[LIM 2] Lima, E., Orientability of smooth hypersurfaces and the Jordan Brower separation theorem, Expo Math. 5 (1987), 283-286.

[MAS] Massey, W., Algebraic Topology, an Introduction, Harcourt Brace, 1968.

[MILN] Milnor, J., Morse Theory, Princeton Press, 1963.

[PIC] Picard, E., Traité d'Analyse, Toure I, Gauthier-Villars, Paris, 3eme ed. 1922.

[WAR] Warner, F., Foundations of Diferentiable Manifolds and Lie Groups, 2nd edition, Springer, 1986.

Index

Universitext

Aksoy, A., Khamsi, M. A.: Methods in Fixed Point Theory
Aupetit, B.: A Primer on Spectral Theory
Bachem, A., Kern, W.: Linear Programming Duality
Benedetti, R., Petronio, C.: Lectures on Hyperbolic Geometry
Berger, M.: Geometry I
Berger, M.: Geometry II
Bliedtner, J., Hansen, W.: Potential Theory
Booss, B., Bleecker, D. D.: Topology and Analysis
Carleson, L., Gamelin, T.: Complex Dynamics
Carmo, M. P. do: Differential Forms and Applications
Cecil, T. E.: Lie Sphere Geometry: With Applications of Submanifolds
Chandrasekharan, K.: Classical Fourier Transforms
Charlap, L. S.: Bieberbach Groups and Flat Manifolds
Chern, S.: Complex Manifolds without Potential Theory
Chorin, A. J., Marsden, J. E.: Mathematical Introduction to Fluid Mechanics
Cohn, H.: A Classical Invitation to Algebraic Numbers and Class Fields
Curtis, M. L.: Abstract Linear Algebra
Curtis, M. L.: Matrix Groups
Dalen, D. van: Logic and Structure
Das, A.: The Special Theory of Relativity: A Mathematical Exposition
Devlin, K. J.: Fundamentals of Contemporary Set Theory
DiBenedetto, E.: Degenerate Parabolic Equations
Dimca, A.: Singularities and Topology of Hypersurfaces
Edwards, R. E.: A Formal Background to Higher Mathematics I a, and I b
Edwards, R. E.: A Formal Background to Higher Mathematics II a, and II b
Emery, M.: Stochastic Calculus in Manifolds
Foulds, L. R.: Graph Theory Applications
Frauenthal, J. C.: Mathematical Modeling in Epidemiology
Fuks, D. B., Rokhlin, V. A.: Beginner's Course in Topology
Gallot, S., Hulin, D., Lafontaine, J.: Riemannian Geometry
Gardiner, C. F.: A First Course in Group Theory
Gårding, L., Tambour, T.: Algebra for Computer Science
Godbillon, C.: Dynamical Systems on Surfaces
Goldblatt, R.: Orthogonality and Spacetime Geometry
Gouvêa, F. Q.: *p*-Adic Numbers
Hahn, A. J.: Quadratic Algebras, Clifford Algebras, and Arithmetic Witt Groups
Hájek, P., Havránek, T.: Mechanizing Hypothesis Formation
Hlawka, E., Schoißengeier, J., Taschner, R.: Geometric and Analytic Number Theory
Holmgren, R. A.: A First Course in Discrete Dynamical Systems
Howe, R., Tan, E. Ch.: Non-Abelian Harmonic Analysis
Humi, M., Miller, W.: Second Course in Ordinary Differential Equations
for Scientists and Engineers
Hurwitz, A., Kritikos, N.: Lectures on Number Theory
Iversen, B.: Cohomology of Sheaves
Kelly, P., Matthews, G.: The Non-Euclidean Hyperbolic Plane
Kempf, G.: Complex Abelian Varieties and Theta Functions

Universitext

Kloeden, P. E., Platen E., Schurz, H.: Numerical Solution of SDE Through Computer Experiments
Kostrikin, A. I.: Introduction to Algebra
Krasnoselskii, M. A., Pokrovskii, A. V.: Systems with Hysteresis
Luecking, D. H., Rubel, L. A.: Complex Analysis. A Functional Analysis Approach
Ma, Zhi-Ming, Roeckner, M.: Dirichlet Forms
Mac Lane, S., Moerdijk, I.: Sheaves in Geometry and Logic
Marcus, D. A.: Number Fields
McCarthy, P. J.: Introduction to Arithmetical Functions
Meyer, R. M.: Essential Mathematics for Applied Fields
Meyer-Nieberg, P.: Banach Lattices
Mines, R., Richman, F., Ruitenberg, W.: A Course in Constructive Algebra
Moise, E. E.: Introductory Problem Courses in Analysis and Topology
Montesinos, J. M.: Classical Tessellations and Three-Manifolds
Nikulin, V. V., Shafarevich, I. R.: Geometries and Groups
Oden, J. J., Reddy, J. N.: Variational Methods in Theoretical Mechanics
Øksendal, B.: Stochastic Differential Equations
Porter, J. R., Woods, R. G.: Extensions and Absolutes of Hausdorff Spaces
Rees, E. G.: Notes on Geometry
Reisel, R. B.: Elementary Theory of Metric Spaces
Rey, W. J. J.: Introduction to Robust and Quasi-Robust Statistical Methods
Rickart, C. E.: Natural Function Algebras
Rotman, J. J.: Galois Theory
Rybakowski, K. P.: The Homotopy Index and Partial Differential Equations
Samelson, H.: Notes on Lie Algebras
Schiff, J. L.: Normal Families of Analytic and Mesomorphic Functions
Shapiro, J. H.: Composition Operators and Classical Function Theory
Smith, K. T.: Power Series from a Computational Point of View
Smoryński, C.: Logical Number Theory I: An Introduction
Smoryński, C.: Self-Reference and Modal Logic
Stanišić, M. M.: The Mathematical Theory of Turbulence
Stillwell, J.: Geometry of Surfaces
Stroock, D. W.: An Introduction to the Theory of Large Deviations
Sunada, T.: The Fundamental Group and the Laplacian (to appear)
Sunder, V. S.: An Invitation to von Neumann Algebras
Tamme, G.: Introduction to Étale Cohomology
Tondeur, P.: Foliations on Riemannian Manifolds
Verhulst, F.: Nonlinear Differential Equations and Dynamical Systems
Zaanen, A. C.: Continuity, Integration and Fourier Theory